THE NATURAL PHILOSOPHY OF LEIBNIZ

THE UNIVERSITY OF WESTERN ONTARIO SERIES IN PHILOSOPHY OF SCIENCE

A SERIES OF BOOKS IN PHILOSOPHY OF SCIENCE, METHODOLOGY, EPISTEMOLOGY, LOGIC, HISTORY OF SCIENCE, AND RELATED FIELDS

VOLUME 29

THE NATURAL PHILOSOPHY OF LEIBNIZ

Edited by

KATHLEEN OKRUHLIK
Department of Philosophy, University of Western Ontario

and

JAMES ROBERT BROWN
Department of Philosophy, University of Toronto

D. REIDEL PUBLISHING COMPANY

A MEMBER OF THE KLUWER  ACADEMIC PUBLISHERS GROUP

DORDRECHT / BOSTON / LANCASTER / TOKYO

Library of Congress Cataloging in Publication Data

Main entry under title:

The Natural philosophy of Leibniz.

(The University of Western Ontario series in philosophy of science ; v. 29)
"The papers in this volume stem from two conferences held at the University of Western Ontario in the spring of 1982 and at the University of Toronto in the fall of the same year"–Acknowledgements.
Includes index.
1. Leibniz, Gottfried Wilhelm, Freiherr von, 1646–1716–Congresses.
2. Science–Philosophy–Congresses. 3. Physics–Philosophy–Congresses.
I. Okruhlik, Kathleen, 1951– II. Brown, James Robert. III. Series.
Q143.L475N38 1986 501 85-24432
ISBN 90-277-2145-9

Published by D. Reidel Publishing Company,
P.O. Box 17, 3300 AA Dordrecht, Holland.

Sold and distributed in the U.S.A. and Canada
by Kluwer Academic Publishers,
190 Old Derby Street, Hingham, MA 02043, U.S.A.

In all other countries, sold and distributed
by Kluwer Academic Publishers Group,
P.O. Box 322, 3300 AH Dordrecht, Holland.

Printed in The Netherlands

Table of Contents

In memory of
John Vincent Strong, S.J.
1939 - 1979

ACKNOWLEDGMENTS

The papers in this volume stem from two conferences held at the University of Western Ontario in the spring of 1982 and at the University of Toronto in the fall of the same year. We would like to thank the conference participants and especially Maxine Abrams and Howard Duncan. For financial support we thank the two philosophy departments and the Social Sciences and Humanities Research Council of Canada.

Kathleen Okruhlik and James Robert Brown

INTRODUCTION: THE NATURAL PHILOSOPHY OF LEIBNIZ

The scholarship of recent years has recognized and emphasized the many distinct currents that contributed to the Scientific Revolution. It has become commonplace to acknowledge the central role played by Hermeticism, alchemy, neo-platonism, scholastic Aristotelianism, and ancient atomism. Further, there is a growing sophistication in recognizing diversity even within the so-called mechanical tradition itself. In light of this recognition it should not be surprising to discover a great deal of residual tension among the conceptual elements central to the Scientific Revolution. These conceptual tensions are presently being explored in ever greater depth by historians and philosophers of science. There is, in consequence, an increasing retrospective appreciation of the enormous complexity of the issues involved.

But, even in the midst of the intellectual tumult which characterized the seventeenth century, some were keenly sensitive to the conceptual difficulties lurking in the foundations of the new physics. Outstanding among these was Gottfried Wilhelm von Leibniz (1646 - 1716) who served both as a contributor to and a critic of the emerging science. No one saw the need for conceptual clarification and modification better than he. No one struggled harder than he did to provide a coherent metaphysical underpinning for an empirically adequate physics. As historians and philosophers of science come to appreciate more fully the complexity of the Scientific Revolution and the depth of the tensions it embodied, they are coming also to better appreciate Leibniz's criticisms of some of its chief architects. By the same token they are coming to a deeper understanding of some of Leibniz's metaphysical tenets which seem bizarre

K. Okruhlik and J. R. Brown (eds.), The Natural Philosophy of Leibniz, 1–6.

outside the context of the scientific milieu in which they were conceived.

Although much Leibniz scholarship over the years has been of very high quality, there have been some shortcomings. One of these has been the relative lack of attention paid to Leibniz's physics. A possible reason for the unfortunate whiggish tendency to treat Leibniz's dynamics as a mere historical curiosity (especially in the Anglo tradition) is the fact that it was in many respects so quickly superseded by Newton's mechanics. But Leibniz's work in dynamics was valuable in its own right and as a springboard to much of his work in metaphysics. The real problems he faced in mechanics, in optics, and in science generally were for Leibniz very often a source of philosophical problems; and the solutions he found there served as models or archetypes for solutions to philosophical problems which were formally analogous.

A second drawback of traditional Leibniz scholarship has been the emphasis placed on Leibniz's rationalism. This has tended to obscure the robust role Leibniz saw for experience and experimental activities. Moreover, it has seemed to preclude a serious role in science for hypotheses and conjectures, and it has tended to make scholars overlook Leibniz's view of the tentative and fallible nature of human scientific activity.

The relative neglect of Leibniz's work on science proper has not been without broader consequences as well. For one thing, it has made it seem that his scientific writings are of merely peripheral interest to the understanding of his other philosophical work, thus robbing us of a deeper understanding of his metaphysics. Further, it has made it appear that Leibniz had little to say which could be relevant to contemporary problems in the philosophy of science. The papers in this volume attempt to dismantle some of the old myths and to forge a new understanding of Leibniz based on a fuller appreciation of his scientific work. We turn now to a brief account of those papers.

"The Problem of Indiscernibles in Leibniz's 1671 Mechanics" by François Duchesneau takes a first step toward assessing the role played by the principle of indiscernibles in Leibniz's dynamics. In his first mechanics, Leibniz has to reconcile a *phoronomia elementalis*, based on a rational reconstruction of motions from conatus as motion indivisibles, with a physical approach to impact which could avoid the inadequacies of empirical concepts in mechanics. In *Theoria motus abstracti* and *Hypothesis physica nova*, the conciliation is worked out by resorting to auxiliary postulates. A purely geometrical determination of conatus would entail exhaustion in successive impacts, suppressed impenetrability beyond the instant, and cancellation of angular deviations, thus preventing circular motions and conservation of causal efficacy. To make up for such shortcomings, Leibniz combines an ether hypothesis with a metaphysical construction of psychological indivisibles underlying the motion indivisibles. In shifting to his dynamics, Leibniz would replace such auxiliary postulates by architectonic principles, such as the principle of indiscernibles.

In an important and perhaps revolutionary paper, Daniel Garber suggests that the relationship between Leibniz's metaphysics and his physics needs to be re-examined. On the usual analysis the two are reconciled by relegating them to separate ontological domains: metaphysics describes the way things really are, and physics the way things appear to us. In his paper called "Leibniz and the Foundations of Physics: The Middle Years", Garber does not deny that this picture is a largely accurate reflection of the views expressed in letters to de Volder and des Bosses, but he does argue that it is not the *only* picture Leibniz held in his mature writings. Garber's chief claim is that a very different picture emerges if we look at Leibniz's work from a few years earlier, in the mid-1680s and '90s when Leibniz was a practicing physicist as well as a metaphysician. On that picture, Garber suggests, physics is a science ***not*** of appearances and of phenomena, but an account of the laws that govern ***a real world of quasi-Aristotelian substances.*** Physics deals with forces that derive from the souls and bodies of multitudes of tiny organisms and with the laws that God has imprinted on these souls. Furthermore, the physics of aggregates (that is, of bodies of everyday experience) is exactly the same as the physics of these corporeal substances, despite the ontological gap between the two domains. Garber suggests that Leibniz's science "loses its grip on reality" only later in his life, after he has set aside his serious work in physics.

Ian Hacking also proposes a different understanding of Leibniz in his article "Why Motion is Only a Well-Founded Phenomenon". Leibniz held that space and time are only well-founded phenomena, but are not real. He thought the same of motion. Readers have commonly taken the unreality of motion to be a mere corollary of the unreality of space and time. Hacking argues that the opposite is the case. Leibniz deduced the unreality of space and time from the unreality of motion. He had important criteria, based on invariance under transformation, that led him to hold motion to be unreal. But *vis viva*, in contrast, was invariant and therefore real. The paper makes comparisons with Newton's view that centrifugal force is not real, with the Coriolis force (1835) and with Richard Feynman's exposition of Einstein's general theory of relativity.

Graeme Hunter's "Monadic Relations" notes that analysis of the Leibnizian notion of representation reveals how it is possible for relations to depend on qualities, without being reducible to them. This in turn provides insight into Leibniz's views concerning the relation of mind and body, the relation of monad to world, and the causal interaction of substances. Within the framework of the Leibnizian ontology which Hunter argues for, it is shown, on the basis of the concepts "action" and "passion", how Leibniz intended physical and metaphysical theories to fit together.

Robert McRae in "Miracles and Laws" notes that according to Leibniz, the Cartesians' laws of nature and Newton's law of gravitation were really only formulations of perpetual miracles. Because miracles are exceptions to laws the issue becomes one concerning the nature of law. It is a major topic in Leibniz's exchange with Bayle, speaking for the Cartesians, and with Clarke, speaking for Newton. One striking result of McRae's analysis is that Leibniz's conception of the miraculous appears to commit him, contrary to his deepest intentions, to holding that the laws of motion in this world are the same for all possible worlds.

Kathleen Okruhlik's "The Status of Scientific Laws in the Leibnizian System" deals with a very tangled web of Leibnizian theses. According to Leibniz, the laws of nature are absolutely contingent, hypothetically necessary, and in some cases *a priori* deducible. This paper attempts to sort out these claims and to make it clear why Leibniz felt compelled to adopt such a complex position regarding laws of nature. In the process, several other key notions in his natural philosophy are introduced and discussed. These include theses regarding analyticity, compossibility, architectonic principles, and the role of conjecture in scientific practice. The view of scientific methodology which emerges is rather more complex than some traditional accounts might lead one to expect.

Robert Butts's article, "Leibniz on the Side of the Angels", explores some unusual but intriguing Leibnizian themes. It concentrates on the methodological role played by angels in Leibniz's writings. Angels are heuristically useful devices because of their near-perfect epistemic orientation: their perceptions suffer much less than ours from distortion and confusion. Leibniz's belief in the intelligibility of the natural world means that there is a method we humans can employ in order to achieve an explanatory perspective close to that of angels. If angels wished to reveal to us the intelligible causes through which all things come about, these causes would be mechanical and mathematically expressible.

"Leibniz and Kant on Mathematical and Philosophical Knowledge" by Jürgen Mittelstrass closely examines Kant's view on the differing natures of mathematical and philosophical knowledge and compares this view to the Leibnizian position. Kant thought that his insistence on the fundamental difference between the two sorts of knowledge distanced him from the Leibnizian camp. Mittelstrass traces the development of Kant's views on the distinction from the *Prize Essay* of 1764 to the *Critique of Pure Reason.* Then he offers an interpretation of Leibniz's methodology in terms of a process of "reconstruction". He suggests that mathematics and philosophy, if their method were construed as reconstruction, would possess one basis in common. The distinction between philosophical analysis and mathematical construction would then lie merely in the dependence of the concept of construction on conditions of pure intuition. Thus, according to Mittelstrass, if one emphasizes the systematic unity of construction and analysis within the concept of reconstruction, the positions of Kant and Leibniz do not lie too far apart from each other. This conclusion contrasts strongly with Kant's original view.

In his article, "Leibniz's Theory of Time", Richard Arthur argues that Leibniz conceived time principally as a structure of relations among monadic states rather than among phenomena only. By reconstructing the temporal theory in modern set-theoretic terms, Arthur tries to show that his interpretation does not presuppose an absolute or "intra-monadic" time for each monad. Futher he argues that the ideality of time pertains only to time as an abstract entity and in no way precludes the states of actual substances or phenomena from occurring in real temporal relations to each other. Finally, Arthur contends that Leibniz's theory is not a causal theory of time, contrary to popular belief.

In "Leibniz and Scientific Realism", William Seager enlists the aid of Leibniz in order to examine the issue of scientific realism, i.e., whether science's employment of theories which mention unobservables amounts to the postulation of the existence of these things or is merely a useful way to develop greater empirical adequacy. Several Leibnizian insights prove to be particularly useful in illuminating questions recently canvassed in Bas van Fraassen's book, *The Scientific Image.* Ultimately these Leibnizian insights (in tandem with other arguments) lead Seager to reject the anti-realism defended by van Fraassen.

It is hoped that these papers will contribute in a significant way to the ongoing re-assessment of Leibniz's natural philosophy and its relevance to his other philosophical work as well as to issues still current in philosophy of science.

Department of Philosophy
University of Western Ontario

Department of Philosophy
University of Toronto

François Duchesneau

THE PROBLEM OF INDISCERNIBLES IN LEIBNIZ'S 1671 MECHANICS

In order to assess the role played by the principle of indiscernibles in Leibnizian dynamics, one needs to get back to the original theses in *Theoria motus abstracti* and *Hypothesis physica nova* (1671), some of which Leibniz will renounce afterwards. I will try to show that the defective theses create a need for new mechanical hypotheses conforming to the principle. Then, I shall address the question: how does the critical revision of the initial theses determine the role of the principle in raising a more adequate theory? Then, one should attempt to characterize how the principle intervenes in the major expositions of the dynamics. Finally, I should set myself the objective of following the process whereby Leibniz justifies the "epistemological" function of the principle in his correspondence with De Volder. In this essay, I shall limit myself to a critical analysis of the 1671 texts.

I

In *Theoria motus abstacti*, Leibniz's project is to lay the foundations for a *phoronomia elementalis* that will "ad instar geometriae" develop in a rational and abstract way the laws of mechanics. In particular, this *phoronomia* should give the rules of impact their most general expression, apart from those physical phenomena that experience reveals. Rules of impact obtained by contruction from experience do not meet the standard of demonstrative knowledge; for such a knowledge requires deductive arguments based on the axioms of geometry and appropriate definitions, i.e. definitions which do not imply any internal inconsistency in conceiving

K. Okruhlik and J. R. Brown (eds.), The Natural Philosophy of Leibniz, 7–26.

the object defined. This systematic approach can be inferred from the section in *Theoria*, entitled *Usus*, where Leibniz sets his project in contrast with contemporary works in mechanics. What proves interesting in the Leibnizian position is that it is supported by a conception of how explaining the phenomena connects with *a priori* reasons; and this conception breeds problems. In the Galileo-type *phoronomia experimentalis*, arguments are built, if one follows Leibniz, from concepts signifying empirical features: hence unamendable indeterminacies about the nature of the extensive continuum, about the reason for circular motion, about the principle accounting for variation in motions in the cases of impact, about the cause of cohesion in bodies. To get rid of those difficulties, it seems fit to borrow from geometry means of setting unambiguous principles that will elicit mechanics as a system.[1] However, two features of the *phoronomia elementalis* show that for Leibniz the abstract geometry of motion is non-nominalist. First, the theory of abstract motion which will be justified by geometrical reasons, shall apply *ipso facto* to physical reality, because the essence of physical objects only corresponds to the possibility of their being geometrically generated, even if this possibility does not suffice for generating them without a sufficient reason for the systematic order they realize *in concreto*. One can in fact conceive bodies as generated by motion alone in an extension which it shapes into a variety of forms. This shaping up corresponds to the delineation of cohesive parts in extension, and cohesion itself results from the reciprocal motions of integrant parts, provided analysis be pursued down to the level of constituent "indivisibles". The relationship between geometry and physics, and the distinction of both from mechanics (understood along the pattern of *phoronomia experimentalis*), show up clearly in the *Problema generale* section of *Theoria motus abstracti*:

> Omnes possibiles lineas, figuras, corpora et motus secundum omnes lineas physice construere meris motibus rectis inter se aequalibus, item meris motibus curvis cujuscunque generis, adhibitis corporibus quibuscunque. Triplex constructio est: Geometrica, id est imaginaria, sed exacta; Mechanica, id est realis, sed non exacta; et Physica, id est realis et exacta. Geometrica continet modos, quibus corpora construi possunt, licet saepe a solo Deo, dummodo scilicet non implicare intelligantur, ut si circulus fiat flexione rectae per minima; Mechanica nostros; Physica eos, quibus natura res efficere

potest, id est quos corpora producunt se ipsis. (P, IV, 235)

Second, the principle of deductively connecting the theory of abstact motion with the "exact" conception of physical reality is grounded on Leibniz's resorting to Cavalieri's method of *indivisibilia* and on the "realist" interpretation Leibniz gives of the extension and motion indivisibles. This interpretation is moreover connected to the use of the sufficient reason principle, which warrants passing from abstract signification to reference in the system of actualized essences. The section *Fundamenta praedemonstrabilia* details precisely those abstract concepts on which rests the ontological relevance of *phoronomia elementalis*. Inspired by Hobbes, Leibniz's theory of the *conatus* is shown to have features that contrast epistemologically with the parallel theses in *De Corpore*. In particular, Hobbes would hold geometrical analysis of the extensive continuum in terms of conatus to be a purely abstract codification for the system of sensory appearances; thanks to such a codification, one could bring about general computations to be applied thereafter to phenomena in accordance with the geometrical conventions previously agreed on. But Leibniz considers *conatus* as the ingredients of reality, while at the same time he uses them as conceptual tools of a symbolic nature in analyzing the possible modifications of motion.

It is in order to account for the start and end of motions, that space, time and motion indivisibles are presupposed. For space, the indivisible is an *inextensum* endowed with the property of juxtaposing with other *inextensa* in infinite number so as to generate finite extensions. Because extension of this physical point is incommensurable with any assignable size but owns at the same time the ontological characteristic of all effective extension (i.e. bearing juxtaposed parts), its integrant parts are said to be *indistantes*, meaning by that the ultimate character of the division by which such points would be obtainable.

> Punctum non est, cujus pars nulla est, nec cujus pars non consideratur; sed cujus extensio nulla est, seu cujus partes sunt indistantes, cujus magnitudo est inconsiderabilis, inassignabilis, minor quam quae ratione nisi infinita ad aliam sensibilem exponi possit, minor quam quae dari potest; atque hoc est fundamentum Methodi Cavalierianae, quo ejus veritas evidenter demonstratur, ut cogitentur quaedam ut sic dicam rudimenta seu initia linearum figurarumque qualibet dabili minora. (P, IV, 229)

Definition of the point as the inextensive element in extension serves to set up that of the *conatus*, inasmuch as it will be called to play an analogous role in analysis of motion. Cf. prop. 10: "Conatus est ad motum, ut punctum ad spatium, seu ut unum ad infinitum, est enim initium finisque motus." (P, IV, 229). This analogical definition of conatus depends on some premisses in the text. There is an essential heterogeneity between motion and rest, which is symbolized by the ratio of one to zero. This radical distinction entails as a consequence that the continuity of motion cannot be recombined from discrete elements of motion with interpolated intervals of rest: hence Gassendi's theory is defective on the manner motions would combine and relative inertia would explain resistance to motion and possibility of reaction in impacts. In contrast with Gassendi, Leibniz proposes an axiom of inertia sanctioning the specific disparity of motion and rest: "(8) Ubi semel res quieverit, nisi nova motus causa accedat, semper quiescet." "(9) Contra, quod semel movetur, quantum in ipso est, semper movetur eadem velocitate et plaga." This notion of inertia, jointly with conatus theory, involves some of the outstanding consequences in Leibniz's first mechanics: for instance, infinite propagation of motion through space, however sizable the obstacles be. The reason is that the conatus as "action commençante" begins to move the obstacle in totality, even if the resulting effect from conatus as "action évanouissante" is impeded. A plurality of opposite conatus may coexist at the same time in the same body, and ensuing motion will correspond to a mutual compensation between conatus. (The state of rest in a body submitted to impact is to be counted as null in computing the result). In the time interval of one conatus (if one presumes a strict homogeneity of time indivisibles), the intrinsically inassignable space scoured by the point in motion will be so much larger since the elementary conatus is more considerable, and in all events, it will be incommensurable, when compared with the expansionless space held by a mere extensive indivisible. In fact, this mere extensive indivisible gets physical reality only in the shape of a line of expansion issuing from conatus in the instant. For our understanding, the emergence of a mass element would be subordinate to an embryonic motion as a *sine qua non* condition; it comes to the same as if the specific and fundamental concept of mass dissolved in indetermination due to a mutual compensation of contrary conatus in the instant. The coexistence within the homogeneous time indivisible of contrary conatus in a material point probably affords the best analogue of mass, as witness §15 and §16:

> (15) Tempore impulsus, impactus, concursus, duo corporum extrema seu puncta se penetrant, seu sunt in eodem spatii puncto; cum enim concurrentium alterum in alterius locum conetur, incipiet in eo esse, id est incipiet penetrare, vel uniri. Conatus enim est initium, penetratio, unio; sunt igitur in initio unionis, seu eorum termini sunt unum, (16) ergo corpora, quae se premunt vel impellunt, cohaerent: nam eorum termini unum sunt, jam *ὧν τὰ ἔσξατα ἕν*, ea continua seu cohaerentia sunt, etiam Aristotele definiente, quia si duo in uno loco sunt, alterum sine altero impelli non potest. (P, IV, 230)

But a coexistence of conatus does not outstretch the instant; only a non-impeded conatus will generate a motion which will continue beyond the instant by virtue of the principle of inertia. Coexisting conatus will combine their efforts according to a formula of algebraic addition/subtraction. In case of *conatus servabiles* or *componibiles*, the effect is a conjunction of motion which at a maximum is equivalent to the motion which the conatus will effect severally. In the case of *conatus incomponibiles*, the ensuing motion is less than the potential effect of the independent conatus. The result of abatement corresponds to the algebraic difference between the two conatus, with the direction of the stronger motion being kept. The typical case of *conatus servabiles* is afforded by a circle rolling on a straight line in a plane. A point chosen on the circumference describes a cycloid in the plane by combining so to speak a circular with a straight motion: both motions are mixed in a way "per minima seu per conatus" (§19, P, IV, 232, and cf. theorem 7, P, IV, 233, in which Leibniz mentions that two conatus may be added such as the straight and the circular determinations in a spiral, the speed respectively generated being kept). The fact that conatus may be kept beyond the instant pursuing a non-rectilinear determination is connected to a major difficulty in the abstract theory of motion. The motion generated by an instantaneous conatus is rectilinear, is uniform, and proceeds with constant speed, while it is not impeded. Keeping a curvilinear determination implies a different order of cause than what is implied by conatus as indivisibles forming the elements of bodies.

A similar problem is raised when one has to account for accelerated motion. Within the same physical point (defined in terms of conatus), a conatus as a cause intervenes in generating a simultaneous conatus which as an effect, integrates the causal property (whatever that be) of the previous conatus and renders it part of its own causal determination to be exerted on the next embryonic conatus. The cumulative process builds the integral effect of acceleration. The same conversely could be asserted for deceleration. One conatus is "sign" of the next insofar as one is cause and the other effect in the instant; but the second is at the same time an analogous sign of causal power for the next in a row, the series of signs being ampliative *ad infinitum*. Cf. §18: "in motu accelerato, qui cum quolibet instanti atque ita statim ab initio crescat, crescere autem supponat prius et posterius; necesse est et casu in instanti dato signum unum alio prius esse." (P, IV, 230-231). In the signs, i.e., the nominal moments comprised within the instant whose succession is uniform, embryonic conatus present themselves issuing from the same initial conatus, and these conatus presuppose a form of integration not comprised in the essential form of a conatus (according to the theory of abstract motion). Leibniz will get back to that point at the end of *Problemata specialia*: he will then note that abstract reasons do not suffice to give an adequate representation of acceleration-deceleration. Indeed, Leibniz believes he can resolve both types of problem: the production of circular motions and the process of effect integration in uniform accelerated motions, by resorting to constructions conformable to the principles of *phoronomia elementalis*, even though the justifications or sufficient reasons for those constructions are not to be deduced from the abstract theory.

For the first type of problem, the phoronomic construction is grounded on the "third way" hypothesis. While discussing impact of equal *conatus incomponibiles*, Leibniz suggests it: Cf. §23 "Si conatus incomponibiles sunt aequales, plaga mutuo deceditur, seu tertia intermedia, si qua dari potest, eligitur, servata conatus celeritate..." (P, IV, 232) We shall examine which type of intermediary is alluded to in compliance with the methodology of abstract phoronomical representation, but it is to be noted first that the justification for the abstract construction seems to surpass any purely geometrical symbolization.

> Hic est velut apex rationalitatis in motu, cum non sola subtractione bruta aequalium, sed et electione tertii propioris, mira quadam sed necessaria prudentiae specie res conficiatur, quod non facile alioquin in tota Geometria aut phoronomia occurrit: cum ergo caetera omnia pendeant ex principio illo, totum esse majus parte, quaeque alia sola additione et

> subtractione absolvenda Euclides praefixit Elementis; hoc unum cum fundam. (20) [Corpus quod movetur, sine diminutione motus sui imprimit alteri id, quod alterum recipere potest salvo motu priore] pendet ex nobilissimo illo (24) Nihil est sine ratione, cujus consectaria sunt, quam minimum mutandum, inter contraria medium eligendum, quidvis uni addendum, ne quid alterutri adimatur multaque alia, quae in scientia quoque civili dominatur. (P, IV, 232)

Clearly, this minimal alteration or addition which will make it possible to withhold the motive determinations from cancelling each other, when conatus are geometrically equal but strictly opposite in respective direction, pertains to an architectonic design in physics. This process for preventing mutual destruction of conatus is conceived by supposing that moving points clash at an angle instead of there being a straight line of impact. But one must determine what are those constantly reiterated *clinamina* which save conatus from mutual destruction (by subtraction of equivalents). Theorem 7 brings some clarification to that question. Leibniz points out that the angular impact of *equivelocia* results in the same straight trajectory for both bodies after the encounter (through bisection of the angle). However, the theory of abstract motion entails that in this way angular deviations would fade out more or less rapidly, instead of reiterating. But Leibniz mentions hypothetical cases in which this fading out would not apply: (1) let us suppose the equivalent conatus at the instant of impact are amenable one to a scalar finite measure, the other to a differential vectorial measure, i.e. for example, the motion of one body is uniform constant and the motion of the other uniform accelerated; then parabolas and other such lines are generated (as shown by Hobbes); (2) let us suppose one has to deal with *conatus servabiles*; then, both conatus can coexist in the physical point, while virtually keeping their respective heterogeneous effects and producing a joint continuous effect: "duo conatus sibi mutuo addi possunt, ut rectus circularisque in spiralem, servata singulorum celeritate." (P, IV, 233). It is easy to find that both cases amount to the same, for the only means to escape the homogenizing of motion determinations and thus to keep to a given angular deviation is that non-rectilinear motions actualize (according to the various patterns of conic sections). And so, one comes back to the primary question: how is circular motion possible, and consequently, real?

Indeed, for Leibniz, the method of indivisibles allows assimilating the indecomposable elements of curvature with the infinitely numerous elements of the polygonal perimeter in the circle's quadrature. With the same type of ratio, the element of a tangent becomes identified with the inassignable secant equivalent to this element. And so one gets an analogue of circular curvature in the form of an angular representation. Thereafter, it is possible to generalize to the elements of curvature for the other conic sections. Take for instance the passage in the *Praedemonstrabilia* wherein Leibniz, following Hobbes, expounds a theory of the angle based on the method of indivisibles, an angle being defined as the quantity of a convergence point (*point de concours*): "Tota de Angulis doctrina est de quantitatibus inextensorum" and so as an illustration: "Arcus minor quam dari potest utique chorda sua major est, quanquam haec quoque sit minor, quam quae exponi potest, seu consistat in puncto." (P, IV, 231) The quadrature is then justified because extensive equivalents can be postulated *ad infinitum* if one abstracts from an inachievable equivalence in terms of finite quantities. But, if technically conatus of circular determinations can be represented by inextensive analogues in terms of rectilinearly-bound conatus and if the quadrature of conic sections can be thus obtained, does this apply as well to the inertial effects of conatus, i.e. to the motions resulting from conatus when no obstacle is interposed? The paradox is noted: "Certe, si arcus et chorda inassignabiles coincidunt, idem erit conatus in recta, qui in arcu: conatus enim est in arcu aut recta inassignabile. Jam si conatus idem est, etiam motus in recta et arcu, id est motus circularis et rectus (quia qualis motus coepit continuatur, seu qualis conatus talis motus) idem erit, quod est absurdum" (P, IV, 232-233). Untying the paradox implies that one distinguishes between "inextensive" quantitites of conatus (from this viewpoint, conatus of circular motion can be identified with conatus of rectilinear motion) and the infinite reiteration of these conatus in the extensive order, all obstacles removed. Straight motion is produced from a conatus without further determination. Circular motion depends on a constantly renewed deviation in the motive disposition along the pattern of a specific curve, i.e. according to an inflexion law which the initial conatus cannot maintain beyond the instant. The indefinite expression of this conatus coincides with a tangent to the curve at the corresponding inassignable point (inassignable angle of secant and tangent at that point). The geometrical expression which renders motion intelligible in *phoronomia elementalis* forbids the conception of conatus as generating other effects than a rectilinear expansion along the tangent. Indeed, a circular motion can be recomposed out of an infinite series of successive rectilinear expressions each of which has an inassignable extension, (i.e. of physical points in the shape of indivisibles). But then the inflexion law which this infinite series represents, cannot be conceived, if one tries to understand it as immanent in any element of the series. The inflexion law transcends conatus insofar as they can get a geometric expression. And in this way, one cannot but resort to the principle of indiscernibles in its architectonic function for the theory.

II

Leibniz's solution to the difficulties we mentioned is left defective in the 1671 abstract mechanics. However, he appeals to some substitutes for the principle of indiscernibles, or, if you prefer, to embryonic versions of those postulates which will be at play in the dynamics. Sect. 17 in the *Fundamenta* suggests that the ontology required to keep the system "alive" shall be completed by adding to conatus as physical points another order of elements capable of embodying the representation of law beyond the instant. From these "psychological" elements, so to speak, the correlation would arise between the inflexion law and motion, motion being the proper expression for conatus as physical points:

> Nullus conatus sine motu durat ultra momentum praeterquam in mentibus. Nam quod in momento est conatus, id in tempore motus corporis....Omne enim corpus est mens momentanea, seu carens recordatione, quia conatum simul suum et alienum contrarium (duobus enim, actione et reactione, seu comparatione ac proinde harmonia, ad sensum et sine quibus sensus nullus est, voluptatem vel dolorem opus est) non retinet ultra momentum: ergo caret memoria, caret sensu actionum passionumque suarum, caret cogitatione. (P, IV, 230)

In short, the suggested solution develops as follows. There is continuous gradation between conatus as ultimate elements of physical reality and selves as elements of substantive individuality expressing themselves through the modalities of phenomenal self-awareness. As a non-constrained effect of conatus, motion has to be rectilinear, uniform, and constant. Circular determination, as well as acceleration-deceleration in rectilinear motions, requires a law to account for the serial and progressive unfolding of states. How this law is to be inserted in the instantaneous conatus cannot be accounted for except on the analogy of a conscious representation spreading over an indefinite sum of time indivisibles: this is the analogy of conscious reminiscence. But the implications of continuous gradation in the analogy do not stop here. The conatus itself is portrayed as a *mens momentanea*. Under the metaphor, an effective difficulty in Leibniz's first mechanics transpires concerning the genesis of *conatus*. Indeed, the problem could lead to developments in metaphysics, as has been the case with some trends towards monadology as a system of monads. But there is an epistemological side to the question. The conatus is, geometrically speaking, a mere "motion indivisible". The inertial effect which it issues in the instant as a rectilinear and uniform motion, expresses a "power of expansion" that one cannot succeed in connecting with a concept of mass which is not really to be found in *phoronomia elementalis*. Cohesion in bodies itself depends on a mutual convergence of elementary conatus constantly recreated and joining in a reactive disposition to surrounding pressure. The "stable" system of endogenous conatus forms a counterpart to exogenous conatus. In last resort, the problem boils down to conceiving adequately how conatus penetrate each other in the instant, each forming one of the indistant terms within an indivisible representing the whole of the embryonic motions involved. The integrant conatus tend to destroy beyond the instant by algebraic subtraction. Keeping the whole system of converging conatus as indistant opposite terms within the same extensive indivisible requires a kind of constant actualization which equals a regeneration in time i.e. along the indefinite addition of time indivisibles. Also, in this case, it is implicitly required that a law of serial expression (or expansion) be integrated to the physical point, whence the antitypy of this point would result and form an individuated structure of interactions between a system of the endogenous conatus and a sum of the exogenous impacts. It seems that those rather implicit requirements suggest that the mechanical theory should overcome any indiscernibility in the postulated physical elements. Indiscernibility would infallibly result from a conjunction of factors: (1) exhaustion of antagonistic motive determinations in impact, (2) suppressed antitypy beyond the instant, (3) cancellation of angular deviations in the inertial effect of conatus. Each and all of these factors would entail destruction of an actualized system of conatus; and they form invalidating reasons for any attempt to conceive of physical nature as conforming to self-sufficient laws. To compensate for such systematic gaps, *phoronomia elementalis* needs to supplement determination in the conatus. Either metaphysical sufficient reason will afford this additional determination or one will need auxiliary postulates or hypotheses to secure a straightforward connection between the theory of abstract motion and *phoronomia physica.*

Martial Guéroult has insisted in his book *Leibniz: Dynamique et Métaphysique*[2] on the metaphysical amendments which Leibniz's first mechanics requires. I will only recall some of his main arguments. After showing the difficulties in the conatus theory, Guéroult develops the view that for Leibniz, mind is called upon to play a substitution role for geometrical determinations since it keeps a memory of the various conatus in opposition beyond the mere instant of convergence. One may quote here a fragment of the same period as the *Hypothesis*: "Si corpora sint sine mente, impossible est motum fuisse aeternum [potest diminui sine fine]" (P, VII, 259-260). To account for circular motions, mind is said to keep so to speak, the law of curvature; for body reduced to a physical point would conform to a determination of motion along the tangent. To keep constant the total quantity of velocity in the universe, one must resume postulating that a spiritual entity intervenes. Since the deficiencies in the system of conatus show up in each instant's interactions, one must assume that this action of mind is immanent in the very structures of physical reality: a "lasting" indivisible (of the psychological type) must underlie and determine the instantaneous indivisible (of the purely phoronomic type). A panpsychic metaphysics is thus called forth to fill up the deficit in the geometrical conception of the phenomenal world by means of abstract motions as combinations of conatus. Guéroult concludes: "livré à ses ressources, le mécanisme [leibnizien de 1671] est incapable de fournir le principe (succédané de la masse) grâce auquel l'élément positif de la nature (qu'on l'appelle vitesse, mouvement, force) ne saurait ni diminuer ni augmenter".[3] It is paradoxical to postulate that conatus are conserved and then to apply an algebraic summation (subtraction) of velocities in impact which blots out the original differences in mutual destruction. The non-paradoxical "conservation" of conatus can only be obtained at a level different from that of geometry, namely that of continued mental representations bearing on the laws to be presumed immanent to the order of conatus. The unretrievable discontinuity in the effects would be made up for by the metaphysical postulate of a parallel continuity in conatus endowed with psychological dispositions. Guéroult concludes on this point:

La disparité entre la cause (esprit) et l'effet (mouvement des

> corps) conduit à placer la cause dans une sphère distincte, sans que l'inadéquation soit par là supprimée entre la cause pleine (l'esprit où tout se conserve) et l'effet entier (le mouvement universel qui progressivement diminue ou s'annihile). Il faut donc se contenter d'une affirmation compensatrice: on dit que la cause (l'esprit) restaure et maintient dans l'effet la plénitude originaire; si nous comprenons que l'esprit (qui rend possible le mouvement en enveloppant le passé et l'avenir de sa trajectoire) *doit* évidemment le conserver comme il se conserve lui-même, le comment de cette conservation, c'est-à-dire l'explication physique nous fait défaut." (p. 19)

With Guéroult's analysis has everything been said on the shortcomings of *phoronomia elementalis*? Indeed, a compensation strategy has been shown: Leibniz postulates psychological entities as well as regulatory principles to express the laws pertaining to the functional dispositions of those entities: this world, governed by regulatory dispositions of a general kind, forms the counterpart of the insufficient system of physical conatus. But, precisely since it is an intelligible counterpart, one might look at the very level of the physical theory for models that ensure some amendment to abstract mechanics. This role is played by what I will call "auxiliary postulates". In the first place, there is a requirement to build the representation of the physical system according to the notion of circular motion. One of the more significant statements on that postulate is the 1671 fragment already mentioned: "Si materia prima moveatur uno modo, id est in lineis parallelis, quiescere, et per consequens nihil esse. Omnia esse plena, quia materia prima et spatium idem est. Ergo omnem motum esse circularem aut ex circularibus compositum aut saltem in se redeuntem. Plures circulationes sibi invicem obstare, seu in se invicem agere. Plures circulationes conare coire in unam, seu corpora omnia tendere ad quietem, id est annihilationem" (P, VII, 259-260). Because of the equivocation of space and matter endowed with impenetrability (theory of voidless extension), diffusion of motion through space entails a circular interaction of material parts, and consequently, physical motion can only fit into a closed trajectory (with circle as the norm). Such a conception of physical reality entails in addition that circular motions are nested in one another down to the physical point, i.e. to the primary structure of conatus. Evidently, circular motion, even though it be issued from a law immanent to the conatus, tends to wear out in impacts when meeting with the motions issued from opposite conatus. Nevertheless, it represents a geometrical pattern more in agreement with the requirement that physical points own an immanent determination and a relative autonomy. A. Hannequin in particular has noted the importance of this auxiliary postulate in setting up the system of physical conatus so as to produce an orderly universe:

> Il fallait...que le monde fût plein, ou que tous les points de l'espace fussent sans exception le lieu de conatus appelés à modifier le mouvement qui les traverse, puis qu'il y eût dans le monde une telle économie que le concours incessant de tous ces conatus, bien que restant soumis aux lois géométriques, maintînt dans les corps non seulement une diversité, mais encore une unité qui y fussent le symbole de la diversité et de l'harmonie des conatus des âmes. Il fallait, en un mot, non seulement que le monde, en tant qu'il est le lieu d'une telle économie, fût l'oeuvre d'une intelligence suprême, mais que tous les mouvements y fussent circulaires; et comme on peut concevoir que de tels mouvements s'enveloppent et se multiplient à l'infini dans l'espace, ainsi que font les bulles de la Théorie du mouvement concret, c'est en eux que la multitude également infinie et hiérarchisée des esprits pouvait enfin trouver son expression corporelle. De là vient... que Leibniz fixait le siège de l'esprit non seulement dans un point, mais comme il l'a plus d'une fois répété, dans le centre d'un cercle.[4]

In curvilinear expansion, the immanent determination, or conatus, of physical points expresses itself. As noted, this type of expansion requires a formal determination in each element of trajectory which the moving point successively actualizes. In concrete physics, which conforms to *phoronomia elementalis*, spherical structures will be postulated, nested in one another ad infinitum and modifiable by impact from subtle bodies which penetrate them on various angles. Those *bullae* and *bullulae* will stand for structures conditioning the equilibrium and interaction modalities for a system of phenomena subjected to geometrico-mechanical order, and this will include the structure conditions for conserving function dispositions. Modifying such dispositions to circular motion will precisely result in the empirical effects, and those will be analyzed as regular changes that can be seen as congruent to a law of order. However, maintaining such dispositions with enough stability in nature entails a convergence *ad infinitum* in exogenous conatus forming a series of differential impacts and shaping up in a circle. This means that the differentials of angular deviation in the successive tangents integrate so as to represent a joint direction of impact in time. Stability in circular motions, so to speak, implies that conatus as nascent motions bring about such a tendency to orderly deviation that converging circular determinants form in the plenum of phenomenal extension. Sorts of constantly reiterated "clinamina" issuing from tendency indivisibles will generate each other in succession. The summation law in this case will transcend the infinitesimal quantities making for such indivisibles.

In a way, the summation law must be internal to the conatus, or rather since the conatus is blotted off in the instant (apart from an eventual inertial motion which would develop indefinitely), this law must depend on a "substratum" of conatus that can maintain the tendency. At this stage, the doctrine fails to conceive a "physical" ground for the summation law; the dynamics will offer solutions: it will revise the notion of conatus so as to fit it to express the genesis of impetus, and it will read conatus as differential expressions of *vis viva*. And so, the integration of conatus will depend on the individualized relationships between force and mass. To be more exact: the *points de force* will be considered as centres for expanding motions, that will create infinitely varied effects in a phenomenal homogeneous and undifferentiated space. Hence, the *points de force* will be the physical sources for all differentiation in the order of extension. Conatus, *qua* motion differentials, are abstract and uniform elements in our account of physical action. They form the basic symbols in a general system for expressing tendencies to motion. The symbols refer, however, to concrete elements endowed with irreducible internal diversity. It is from such concrete elements that the diversified tendencies to motion radiate. To an understanding capable of developing an analysis of the "contents" of physical points, they would show up as intrinsically discernible. In Leibniz's final philosophy, a relation of analogical expression will link the strictly individualized monads and the centres of force as sources for heterogeneous motive dispositions. In 1671, the theory does not yet account for the inherence of a law of serial development of conatus in the physical point. However, the circular "enveloppement à l'infini" of motive determinations, therefore of effects issuing from conatus, gives a kind of anticipation of the later theory. At the same time, the inadequacy of the concepts emphasizes the need to add a system of postulates, even though the *phoronomia elementalis* would have abstained from that move. An adequate system of physical postulates should prevent that functional indiscernibility resulting from the notion of equivalent and opposite conatus which annul in algebraic subtraction. The progressive erasing of homogeneous conatus could not be made consistent with the persisting power of generating action in nature.

Similarly, Leibniz's 1671 mechanics lacks a notion of mass ever so little distinct from that shaped extension which delineation of bodies reduces to. One of the consequences would be total dispersion of rectilinear motion in the plenum. Among Leibniz's corrective auxiliary postulates, the theory of the ether plays a special role: it helps account for elasticity in bodies. Indeed, passing from abstract mechanics to the organized system of physical phenomena, one has to suppose a divine arrangement to keep the effects in the most orderly and convenient connection. For instance, in the abstract theory, in all cases of oblique incidence, reflection conforms to a different angular ratio according to the measure of the incidence angle: the smaller angle is complementary (*supplementum ad rectum*) to two times the larger one: equality between incidence and reflection angles is then only a "cas-limite" (when the linear determinations of the incidence angle measure 30 degrees). But experience manifests equality between incidence and reflection angles to be the general rule, and this affords a basis for explaining, for instance, optical phenomena by a sufficient reason of teleological nature (cf. easiest path principle).

> Supposez appliquées par examples dans toute leur nudité et sans modification les lois abstraites de la réflexion, selon lesquelles les angles d'incidence et de réflexion sont inégaux, sauf le cas particulier de l'incidence de 30°, et de là allait suivre que non seulement ni la vue ni l'ouïe ne sauraient exister (visus auditusque existere non posse [P, IV, 187]), mais qu'à l'ordre du monde il manquerait quelque chose, dès lors qu'il échapperait par exemple à cette loi, selon laquelle la nature suit toujours les voies les plus aisées dans la production de ses effets, et de laquelle le géomètre déduit directement les lois de la réflexion et de la réfraction.[5]

On the other hand, this type of corrective to the abstract rules of impact serves in warranting that a given angular deviation in motion reiterate, which is one of the conditions for keeping a curvilinear determination. But one cannot derive from the abstract laws of motion elasticity, which results precisely from such a structure of physical effects.

For want of access to the divine solution, we have put forth an imaginary solution, "hypothèse unique qui nous permette de retrouver la solution divine".[6] Working up this hypothesis is the proper objective of *Theoria motus concreti* (or *Hypothesis physica nova*). The hypothesis itself is a reply to the 11th special problem in *Theoria motus abstacti* which is to find out how bodies can mutually reverberate in impact. Among the theorems which are instrumental in the analysis, let us mention prop. 21: *Corpus discontiguum plus resistit contiguo* (P, IV, 234). A discontiguous body is a provisional reconstruction for an equivalent of mass. If there are innumerable discontiguous parts inside a solid, any exogenous conatus applied to the solid will determine series of impacts in it. Due to the multitude of discontiguous elements in each direction the motive action can in a way dissolve inside the solid without evincing any displacement of the whole. In effect, this theorem cannot be completely demonstrated within the abstract theory. One might even state that it is a paralogism. Difficulties are as follows. The corpuscles located on the line of impact can have no action contrariwise to exogenous impact, except if they exert counter-current conatus, even if minimal. Discontinuity as a geometrical pattern for bodily components is not a condition sufficient to account for the antagonistic determination in conatus. Besides, discontinuity itself seems to entail an additional difficulty pertaining to the interval between parts. Have we not to postulate an intercorpuscular vacuum to ground this structural discontinuity? But the hypothesis of a vacuum has to be excluded. The only way out of the deadlock is to admit (1) that the discontinuity among internal parts is due to motions segmenting the apparent continuum; (2) that those motions are curvilinear and closed. Cf. Th. 22: "Si non datur vacuum, nullus quoque motus rectilineus, aliusve in se non rediens (v.g. spiralis) dabitur. Hinc multa motus in pleno mira consectaria deduci possunt." (P, IV, 234). The theory of the ether will serve to express conditions for the discontiguity of extensive elements in actualized nature. The main property of this subtle fluid is that its parts will not obstruct one another; it is as if there were elements of given constant conatus which intervened only by "reacting" against the motive expansion of some extensive element of a different kind; "reacting" means reverting or inflecting this latter element's conatus without any decrease in speed. This is the true source for the maintained circularity in motion of extensive parts. This is also the sufficient condition for elasticity, as a property of extensive parts. Cf. *Problemata specialia* 11: "Repercussionem mutuam efficere; id fiat, si ambo ferantur a liquido quodam discontiguo propter theor. 21 ita subtili, ut plurimum alterius per alterius polos mutuo, non obstante occursu progediatur; tunc etiam in corpus oppositum impetum mutuo transferent, unde non tantum repercussio, sed et viarum et celeritatum permutatio orietur (P, IV, 235). It is noteworthy that starting from that auxiliary hypothesis, Leibniz will attempt to explain "all observable motions" (*phoronomia physicalis*), and generically, all phenomena of the real world, in particular, those relating to light. The equivalence among kinetic hypotheses (as established by Descartes) will serve to suppose ether moving and bodies moved and modified by the action of the subtle fluid. This is the object of *Hypothesis physica nova.* For our present purpose, let us only note that some mechanical phenomena deemed paradoxical in pursuance of *phoronomia elementalis,* tend to corroborate the theory when it is strengthened by the auxiliary postulate. And so Leibniz pretends to annex mechanical conceptions alien to the abstract system: the Wren-Huygens laws of impact, Descartes's conservation of quantity of motion, the primordial state of elasticity in bodies, the possibility of plain circular motion (cf. Hobbes), material cohesion, etc. In fact, ether as a subtle and penetrating fluid, interferes with the system reduced to geometrically analyzable properties so as to enable the conatus of the various physical points (1) to conform to a law of alternative expansion-compression (allowing for elasticity); (2) to generate circular effects, the irradiating motion issued from conatus being constrained by ether along a regular *clinamen.* (More precisely, this clinamen results from the original determination in the conatus when related to the extensive structure of particles in the surrounding fluid). So, ether in its reactive role serves to circumscribe the sphere of action intrinsic to physical points. As a theoretical entity, its significance consists in this compensatory role.

For sure, the shortcomings of *phoronomia elementalis* are not made up for by mere resorting to the auxiliary postulate. But this appeal shows that the conatus theory is subject to an architectonic rule, that of the radical discernibility of physical points. As radiation centres for conatus, they must retain their prerogative of being physical causal agents. Therefore, they must escape the blotting out which threatens conatus in their rectilinear and uniform expression. In relation to each other and taken in the instant, *conatus incomponibiles* are apt to oppose and dissolve their respective effects in an algebraic subtraction; *conatus componibiles* or *servabiles* show irreducibly distinct determinations that persist in the global phenomenal effect. Evidently, the ether auxiliary hypothesis allows assigning a characteristic *servabilitas* to all conatus in mutual relationship. Such is the derived means whereby Leibniz grants actualized physical points an individuated causal function which expresses through an ordained series of diversified conatus along the homogeneous series of time indivisibles.

In the revised mechanics, the twofold connection between primitive and derivative force, and at each level between active and passive force, will substitute for the arbitrariness of the auxiliary hypothesis in maintaining the perfect individuation of the "centres de force". The role played by ether in the phoronomic system will be superseded by the concept that each "centre de force" owns an internal limitation on its dynamic disposition. This limitation at the phenomenal level is expressed as the reciprocal conditions of elasticity between finite masses which mutually communicate motion. At the substantive level, it expresses the expansion-compression law specific to a physical point in virtue of the potential integration of the diversified resulting conatus.

However, replacement of the auxiliary postulate will be done in a remarkable manner. The junction between physical hypothesis and *phoronomia elementalis* seemed to result from a kind of "deus ex machina", since the two distinct levels of analysis were merely juxtaposed. The ether hypothesis will survive in Leibniz's later physical thinking: he will not really renounce the arguments in *Theoria motus concreti*, while *Theoria motus abstracti* has been replaced by the principles of a new mechanics. But the new theoretical postulates (defining the level of mechanical analysis) do this (ideally) without integrating propositions which would have been derived *a posteriori* from the physical representation of phenomena. The ether theory does not come in anymore to correct the inadequacy of the principles of mechanics. It will only come in as a kind of correspondence rule connecting the formal dynamical concept of elasticity with concepts having empirical referents and signifying the phenomenal effects of elasticity: gravity, magnetic impulse, reflection and refraction. Is not this abstention from appealing to auxiliary hypotheses dependent upon *a posteriori* inferences more programmatic than effective? The question comes to the fore for anyone attempting to analyze the structure of the revised mechanics. Leibniz meets two orders of difficulty in trying to work out his dynamic theory. On the one hand, he must provide a non-defective system of concepts and postulates for the mechanical theory in order to make for the shortcomings of his *phoronomia elementalis*: let us take those difficulties as pertaining to formal coherence within the theory. Second, it is rather manifest that in mechanics purely geometrical analysis should be revised by resorting to architectonic principles, and among these, essentially by resorting to the principle of indiscernibles. The difficulties to overcome here concern the objective significance of the theory; i.e., the translation system to be built between the postulates and those empirical truths which abstract enunciations serve to interpret. In this regard, the theoretical statements must set up the coherence of the phenomenal manifold. If one takes into account the Leibnizian twofold criterion of validity, i.e., analytic and heuristical, for hypotheses, a theory will be worth its power to analytically express more and more developed rational connections among the relevant phenomena.[7] I believe that architectonic principles are not only essential to overcome the second type of difficulties but that they interfere considerably with Leibniz's attempt to perfect his mechanical theory at the formal level. I hope my analysis of the 1671 *Theoria motus abstacti* has been instrumental in supporting this view.

Départment de philosophie
Université de Montréal

Notes

Abbreviation: P for *Leibniz: Die Philosophischen Schriften*, vols. I-VII, edited by Gerhard, (Berlin, 1875-90).

1. I would tend to say that with such an aim in view, Leibniz projects implicitly the notion of an analytical mechanics, which will get its autonomy from physics only with d'Alembert and Lagrange.
2. Paris: Aubier Montaigne, 1967.
3. *ibid.*, p. 17.
4. Arthur Hannequin, "La première philosophie de Leibniz", in *Etudes d'histoire des sciences et d'histoire de la philosophie*, Paris: F. Alcan, 1908, t. 11, p. 162.

5. *ibid.*, p. 70.
6. *ibid.*, p. 72.
7. Cf. Francois Duchesneau, "Hypothèses et finalité dan la science leibnizienne", *Studia Leibnitiana, 12* (1980), p. 161-178; "Leibniz et les hypothèses de physique", *Philosophiques, 9* (1982), p. 223-238.

Daniel Garber

LEIBNIZ AND THE FOUNDATIONS OF PHYSICS: THE MIDDLE YEARS*

Leibniz must appear as something of a paradox to the reader of the recent literature on his thought (i.e., that written after the seminal work of Russell and Couturat). The Leibniz who appears in the commentaries is almost invariably Leibniz the logician/metaphysician, concerned to argue for a world of individual substances, later monads, mind-like, immortal, containing all and reflecting all, concerned to argue that this is all there is to our world and to every other possible world. But, on the other hand, most of us know, if only dimly, that Leibniz was a physicist of some note in his day, and, as such, was concerned with the determination of the basic laws that govern the bodies of everyday experience, the same problem that worried Galileo, Descartes, Huygens, and Newton. But what status could the science of physics possibly have for a philosopher who, like Leibniz, seems to hold a metaphysics so distant from our common-sense conceptions of the world of physics? The perplexity is greater still

*Earlier and much abbreviated versions of this paper were read at the University of Toronto, Ohio State University, and at a joint meeting of the History of Science Society and the Philosophy of Science Association. I would like to thank audiences there for a series of lively discussions. Much of the material in the paper was developed in the course of a seminar I gave on Leibniz at Princeton University in Fall 1982, and I would like to thank the students, faculty, and philosophers from the larger community who attended and made the seminar so valuable for me. And finally, I would like to give special thanks to J.E. McGuire, Leonard Linsky, Howard Stein, and Robert Sleigh for all of their help. This paper is dedicated to H.L.G.-P., who was born at almost the same time as the paper was.

K. Okruhlik and J. R. Brown (eds.), The Natural Philosophy of Leibniz, 27–130.

when one finds, even in the standard works of Leibniz's metaphysical corpus, like the *Discourse on Metaphysics*, numerous remarks to the effect that his metaphysics is intended to *ground* the true physics! What possible connection could there be between Leibniz's metaphysical conception of what there really is in the world, and his physics?

There is something of a standard answer to this question among those who have raised it. According to this answer, the real world for Leibniz is just the world of mind-like simple substances, monads, and the like. On this view, the bodies of physics (along with their geometrical and dynamical properties) enter *only* as phenomena, one mind's confused perceptions of multitudes of other distinct, mind-like monads.[1] Physics, then, is a science, not of real things, it is claimed, but of phenomena. As Jacques Jalabert put it, "with the Platonic tradition, he [Leibniz] limits the truth of mechanism to the world of appearances."[2] Martial Gueroult takes this even a step further, suggesting that the true physics, the physics that we would pursue if we were capable, is nothing but the monadology itself.[3] In short, Leibniz's metaphysics and physics are reconciled in this way by relegating them to separate ontological domains: metaphysics describes the way things really are, and physics, the way things appear to us. As Gueroult puts this view, "...In the universe as the physicist perceives it, God's wisdom does not manifest itself as it is in itself, but, in some sense, for us, that is, through the imperfect and confused vision of a monad."[4]

This general picture of the relation between Leibniz's physics and his metaphysics is right, I think, in outline if not in every detail. But, I suggest, it is not the *only* picture Leibniz held in his mature writings. There is no question but that the standard account is a more or less accurate reflection of the views Leibniz often gives de Volder and des Bosses in explaining the foundations of his physics to them. But when we look at Leibniz's writings from a few years earlier, in the mid-1680's and '90's when Leibniz was both a practicing physicist and metaphysician, a very different picture emerges. On that picture, I claim, physics is a science *not* of appearance and of phenomena; in the era of the *Discourse* and the *Specimen Dynamicum*, physics is, I claim, an account of the laws that govern *a real world of quasi-Aristotelian substances*.

In this essay I would like to discuss this largely unnoticed conception of the metaphysical foundations of Leibnizian physics. I shall begin by outlining at some length an account of the created world that Leibniz gives in his middle years (§§I-IV). I shall then try to relate this non-monadological conception of the world to the notions of force and law, the basic concepts of Leibniz's physics, and I shall try to establish that the physics connected with this, Leibniz's first mature metaphysical system, is firmly grounded in his metaphysical picture of the way the world really is (§§V-VI).

Before entering into the details of this project, though, one remark is in order. Despite the length of this paper, I regard it as work in progress, a very preliminary sketch of a large scale reinterpretation of some texts of Leibniz's that, I think, have not yet been fully understood. While I have tried to deal with the most important questions and objections that can be raised in connection with the reading I propose, I am painfully aware that many details remain to be filled in and that much work is left to be done to make my reading of Leibniz's thought fully convincing. I hope, of course, that this paper will cause others to see in Leibniz what I have seen. But I will be sufficiently gratified if I can lead others to reread familiar texts from a different point of view and rethink comfortable interpretations taken for granted.

I. The Metaphysics of Corporeal Substance: a Preliminary Sketch

I would like to begin this project by briefly sketching out a conception Leibniz had of the world of created things as it appears in some of his writings from the mid-1680's through to the end of the Century, especially the version that he gives in the important *Correspondence with Arnauld* and related documents. Leibniz's view in these writings is strangely unfamiliar, even to Leibniz scholars. Everyone knows about the worlds of the *Monadology*, world upon possible world of mind-like monads, simple substances freely floating in non-space, revelling in their perceptions and appetites, windowless and unconnected by anything at all, with the possible exception of substantial chains. But the world of the 1680's and '90's and its peculiar charms has been almost entirely overlooked.[5] In this period, what is generally acknowledged as the starting place of Leibniz's mature thought, we find, I think, not a world composed of souls alone, but a world whose principal inhabitants are corporeal substances understood on an Aristotelian model as unities of form and matter, organisms or a rudimentary sort, big bugs which contain smaller bugs, which contain smaller bugs still, and all the way down.

One thing that may explain our neglect of this aspect of Leibniz's thought is the fact that this more Aristotelian picture is only barely suggested in the *Discourse on Metaphysics* (DM), what has become for us the canonical text from this period of Leibniz's work. The main focus of the DM is the individual substance.[6] The individual substance is given an abstract characterization in the often cited DM8, and its principal properties are delineated in succeeding sections. Individual substances must all differ from one another, they must be indestructible and indivisible, they must all mirror the universe as a whole, and cannot causally interact with one another, in metaphysical rigor. This much is clear. But Leibniz is not very good with examples in the DM; though we know, in the abstract, what substances are, like the logical atomist's notorious simples, it is very difficult to get our hands on a Leibnizian individual substance *in concreto*. Leibniz seems to agree with Descartes that rational souls count as substances, immaterial or incorporeal substances.[7] He also seems to recognize other kinds of immaterial substances, what he calls, after the Scholastics, substantial forms, entities which are "related to souls," but which lack reason and self-awareness.[8] However, *nowhere* in the DM is it ever asserted that such *souls* or *forms* are *all* that there is to the world. Leibniz, to be sure, rejects the Cartesian conception of corporeal substance, substance that is essentially extended and nothing more.[9] But even though Cartesian bodies are not substances, there is at least the hint that in the Leibnizian world of individual substances, there may be, in addition to the immaterial souls that lack bodies, other substances that have something of the corporeal. So, for example, though Leibniz rejects the Cartesian conception of material substance, his point is *not* that there is *nothing* substantial in extended things. Rather, the claim is that "the *whole nature* of body does not consist in extension, that is, in size, shape, and motion." Leibniz seems to suggest that body must contain, in *addition* to something connected with extension, "something related to souls, which one commonly calls substantial form."[10] And, even more suggestively, in DM34 Leibniz advances the assumption that "bodies which make up an *unum per se*, like human beings, are substances, and have substantial forms," an assumption that he is even willing to extend to beasts.[11] It seems no accident that the principal example Leibniz gives of an individual substance in the crucial DM8 is Alexander the Great, not his soul (which explicitly comes upon the final sentence of that section) but, it seems, Alexander himself, body and soul. There is, then, a suggestion that there *may* be individual substances which, though they may *have* souls or soul-like substantial forms, have bodies as well.

These passages are, admittedly, inconclusive. But these hints Leibniz gives in the DM are fully borne out by the almost contemporary discussion in the *Correspondence with Arnauld* (CA), a document that Leibniz took seriously enough as a representation of his views to consider publishing once, perhaps twice in later years, unlike the DM itself.[12] The *unum per se*, soul united to body, suggested in DM34, is a central focus of Leibniz's later letters with Arnauld. Such substances are what Leibniz there dubs corporeal substances (presumably to distinguish them from rational souls and forms, which are *in*corporeal substances), and are clearly considered full-fledged instances of the more general notion of an individual substance.[13]

Leibniz's main motivation for introducing such corporeal substances in the CA is an argument not found in the DM, an argument that he repeats over and over to Arnauld, an argument that I shall call the *aggregate argument*. Leibniz writes:

> I believe that *where there are only entities through aggregation, there will not even be real entities*; for every entity through aggregation presupposes entities endowed with a true unity....I do not grant that there are only aggregates of substances. If there are aggregates of substances, there must also be genuine substances from which all the aggregates result. One must necessarily arrive either at mathematical points from which certain authors make up extension, or at Epicurus's and M. Cordemoy's atoms (which you, like me, dismiss), or else one must acknowledge that no reality can be found in bodies, or finally one must recognize certain substances in them that possess a true unity.[14]

The claim is a rather simple one: the reality that an aggregate of individuals has derives from the reality of its parts. To use an example Leibniz often appeals to, a pile of stones can only be real if the stones of which it is composed are real. As Leibniz puts it, "I deduce that many entities do not exist where there is not a single one that is genuinely an entity and that every multiplicity presupposes unity."[15] Leibniz takes this general argument to have an obvious application to bodies. If we conceive of extended bodies, as the Cartesians argued, as indefinitely and arbitrarily divisible, as containing extended parts which, in turn, contain smaller extended parts, *ad infinitum*, then it follows that bodies must themselves have no reality (in a sense we shall later discuss). Leibniz writes:

> Now, each extended mass [*masse étendue*] can be considered as composed of two or a thousand others; there exists only an extension achieved through contiguity. Thus one will never find a body of which one can say that it is truly a substance. It will always be an aggregate of many. Or rather, it will not be a real entity, since the parts making it up are subject to the same difficulty, and since one never arrives at any real entity, because entities made up by aggregation have only as much reality as exists in their constituent parts.[16]

But *if* a body is to be real, if, as he puts it, body is to be more than "a phenomenon, lacking all reality as would a coherent dream,"[17] then "one must recognize certain substances [in bodies] that possess a true unity."[18] These are what Leibniz calls corporeal substances. Leibniz is at first tentative in inferring the existence of corporeal substances from the aggregate argument, suggesting that we are only entitled to them under the *assumption* that there is, indeed, something real in bodies, an assumption about which we cannot be absolutely certain.[19] But as the argument progresses, Leibniz becomes more and more confident of the conclusion that bodies must contain something substantial. When Arnauld, for example, questions the necessity for unities in bodies,[20] Leibniz replies:

> You say you do not see what leads me to admit these substantial forms or rather these corporeal substances endowed with true unity; but it is because I cannot conceive of any reality without true unity.[21]

The implication here, and throughout these later letters, is that such a conception of the world of bodies is just not credible. Though Leibniz never claims to have a demonstrative argument that will establish the real existence of the world of bodies beyond all doubt, he never takes seriously the possibility that the world of bodies is unreal. And if there is to be something real in the world of bodies, then there must be something in body that is, unlike bare extension, a genuine unity.

But what, *in concreto*, are the unities that Leibniz has in mind? A passage I quoted earlier makes clear that neither mathematical points nor atoms will do for Leibniz;[22] Leibniz wants something more, something substantial. Other passages suggest that this something substantial essentially involves mind, soul, or form. Leibniz writes, for example:

> If the body is a substance and not a simple phenomenon like the rainbow, nor an entity united by accident or by aggregation like a heap of stones, it cannot consist of extension, and one must necessarily conceive of something there that one calls substantial form, and which corresponds in a way to the soul.[23]

Or, even more explicitly, Leibniz writes:

> ...The substance of a body, if bodies have one, must be indivisible; whether it is called soul or form does not concern me.[24]

These passages suggest that the unities that ultimately ground the reality of bodies, the genuine substances into which the aggregate is resolved, are merely minds, forms, souls, the incorporeal substances we discussed earlier. This, in turn, suggests a view often attributed to Leibniz whereby bodies are simply aggregates of mindlike monads, families of tiny souls, to use Russell's colorful phrase.[25] But despite the obvious suggestion, this cannot be precisely what is going on here. Arnauld at one point argues, appealing to the authority of St. Augustine, that true unity may well be lacking in body, and found only in spirit. He writes:

> ...St. Augustine feels no difficulties about recognizing that bodies possess no true unity, because unity must be indivisible and no body is indivisible. Hence there is no true unity except in spirits, anymore than there is a true 'self' [except in spirits].[26]

Leibniz's answer is instructive. He does *not* reply that the unities that make up bodies are themselves minds. Rather, he replies:

> You object, Sir, that it may be of the essence of body to be devoid of true unity; but it will then be of the essence of body to be a phenomenon, lacking all reality as would a coherent dream.[27]

Whatever these unities are that make up real bodies, they do not seem to be souls *simpliciter*. But, then, what are they?[28] Leibniz's conception of unity as it pertains to the substances that make up real bodies is, perhaps, best appreciated in contrast with cases in which the appropriate sort of unity is lacking. Consider, first, an illuminating example Leibniz gives

Arnauld where, he claims, we have no unity and thus no substance:

> Let us assume that there are two stones, for instance the diamonds of the Grand Duke and of the Grand Mogul: one and the same collective name may be given to account for both, and it may be said that they are a pair of diamonds, although they are to be found a long way away from each other; but it will not be said that these two diamonds compose one substance. Matters of degree have no place here. If therefore they are brought closer to one another, even to the point of contact, they will not be more substantially united on that account; and even if after contact one were to add some other body calculated to prevent their separating, for example, if one were to set them in a single ring, all that will make only what is called *unum per accidens*. For it is as though by accident that they are forced into one and the same movement.[29]

In this case, no corporeal glue, however strong, could make the two diamonds into a single substance. Leibniz continues this passage a few lines later by suggesting that somehow or another the appeal to a soul or form is in order, if we are to have the genuine unity a substance requires:

> Substantial unity requires a complete, indivisible and naturally indestructible entity...which cannot be found in shape or motion...but in a soul or substantial form on the example of what one calls 'self'.[30]

But what does Leibniz have in mind here? Though the passage suggests, once again, that the ultimate unities are just souls, what Leibniz has in mind is something quite different, I think. To see how the soul or form brings about unity, let us turn briefly to Leibniz's account of how the soul produces unity in human beings. In human beings, Leibniz thinks, "the soul is truly the substantial form of our body."[31] Or, as Leibniz tells Arnauld at somewhat greater length:

> ...man...is an entity endowed with a genuine unity conferred on him by his soul, notwithstanding the fact that the mass of his body is divided into organs, vessels, humors, spirits...[32]

An obvious suggestion is that the human body, despite its complex parts, is unified and enters into a genuine substance by virtue of the fact that it is appropriately connected to an immaterial substance, a soul: it is in this way, it seems, that the soul brings about unity; it is in this way that, for human beings, at least, "substantial unity requires...a soul or substantial form".[33] The soul is, as it were, a kind of *incorporeal* glue that unites the different parts of the body and makes them all belong to one genuine individual, one genuine substance. And, consequently, a corpse, a human body not so connected with a soul cannot be a substance, properly speaking, as Leibniz tells Arnauld.[34] But, Leibniz tells Arnauld,[35] it is similar in the world of non-human corporeal substances. That is, the substantial form can provide a non-human body with substantial unity in just the way our soul does so for us, by being appropriately connected to that body. Returning now to the discussion of the two diamonds, Leibniz concludes:

> I accord substantial forms to all corporeal substances that are more than mechanically united....If I am asked for my views in particular on the sun,...the earth, the moon, trees and similar bodies, and even on animals, I cannot declare with absolute certainty if they are animate or at least if they are substances or even if they are simply machines or aggregates of many substances....[E]very part of matter is actually divided into other parts as different as the diamonds [of the Grand Duke and the Grand Mogul]; and since it continues endlessly in this way, *one will never arrive at a thing of which it may be said: 'Here really is an entity,' except when one finds animate machines whose soul or substantial form creates substantial unity independent of the external union of contiguity.* And if there are none, it follows that apart from man there is apparently nothing substantial in the visible world.[36]

So, it seems, corporeal substances, the unities of which the bodies of everyday experience are composed, are to be understood on analogy to human beings, a mind or something mindlike (a substantial form), connected with a body. And this, then, is the proper conclusion of the aggregate argument: for the extended things in the material world to be real, they must, ultimately, be made up of corporeal substances, unities composed of soul and body. Thus while bodies *must* contain something mental or analogous to the mental, a form that will unite discrete bodies and create genuine unity, it is these unities, these corporeal substances, these basic building blocks that ground bodies, and not the incorporeal substances themselves.[37] This resolution of the bodies of everyday experience into fundamental unities, corporeal substances, suggests a kind of atomism, what Leibniz sometimes calls ***metaphysical*** or ***substantial*** atomism to distinguish it from the more familiar Democritean atomism of Cordemoy, for example, a world of basic things distinguished by virtue of their extreme hardness. But there is a crucial difference between Leibniz's substantial atomism and any other atomism current in the 17th century. For the physical atomist, there is a rock-bottom level of analysis; when we divide a body into its ultimate parts, we arrive at atoms, beyond which we cannot go. But not so for Leibniz's metaphysical atomism. When the Leibnizian divides a real body into its ultimate constituents, the corporeal substances, we *can* stop there; that is sufficient to ground the real existence of the body in question. But, it is important to note, we *needn't* stop with the first layer of corporeal substances we come upon. Leibniz's basic building-blocks *themselves* contain *further* corporeal substances, and so on *ad infinitum*. Leibniz writes to Arnauld, implicitly appealing to the so-called principle of plenitude:

> I also believe that to wish to restrict genuine unity or substance to man almost without exception is to be as limited in metaphysics as were in physics those who enclosed the world in a ball. And since genuine substances are as many expressions of the whole universe considered in a certain sense and as many duplications of the works of God, it is in keeping with the greatness and beauty of God's work, since these substances do not impede one another from making as many [substances] in this universe as possible and as higher reasons allow.[38]

These bodies with souls, corporeal substances that, in recognition of the mentality of form, are analogous to animate beings or organisms, are, thus, everywhere in Leibniz's world, in its smallest part. "The whole of matter must be full of substances animate or at least living," Leibniz writes.[39] And thus Leibniz responds to Arnauld's common-sense objection that animate bodies are but a miniscule proportion of the world:

> From that I see, Sir, that I have not yet expressed my ideas clearly so as to make you understand my hypothesis. For apart from the fact that I do not remember saying that there is no substantial form except souls [*ames*], I am very far removed from the belief that animate bodies are only a small part of the others. For I believe rather that everything is full of animate bodies, and to my mind there are incomparably more souls than there are atoms for M. Cordemoy, who makes a finite number

> of them, whereas I maintain that the number of souls or at least of forms is quite infinite, and that since matter is endlessly divisible, one cannot fix on a part so small that there are no animate bodies within, or at least bodies endowed with a basic entelechy or (if you permit one to use the word 'life' so generally) with a vital principle, that is to say corporeal substances, about which it may be said in general of them all that they are living.[40]

Thus, even an individual organism, a corporeal substance, a body united by a form or soul, must *itself* contain other organisms. So, Leibniz writes about human beings, in a passage that we have already seen in part:

> ...man...is an entity endowed with a genuine unity conferred on him by his soul, notwithstanding the fact that the mass of his body [*la masse de son corps*] is dividided into organs, vessels, humours, spirits, *and that the parts are undoubtedly full of an infinite number of other corporeal substances endowed with their own entelechies.*[41]

Or, Leibniz writes about corporeal substances in more general terms:

> If one considers the matter of the corporeal substance not mass without forms but a secondary matter which is the multiplicity of substances of which the mass [*masse*] is that of the entire body, it may be said that these substances are parts of this matter, just as those [substances] which enter into our body form part of it, for as our body is the matter and the soul is the form of our substance, it is the same with other corporeal substances.[42]

With this we have a crude and preliminary outline of the metaphysics of the CA. On this view, the bodies of everyday experience are not made up of Cartesian extended-stuff, but of corporeal substances, soul-like forms united with organic bodies, which organic bodies are in turn made up of smaller corporeal substances, and so on to infinity. This world is, on its surface, at least, very different from the more familiar world of the later *Monadology*. The world of the *Monadology*, it is generally acknowledged, is a world of *souls* (or something analogous) *and souls alone*; everything in the world of the *Monadology* is, ultimately, grounded in the mental. This, though, doesn't *seem* to be what Leibniz has in mind in the CA. One *might*, of course, argue that the *Monadology* lurks beneath the surface of the DM and CA, that a sensitive interpretation of the nuances and details of those writings drives us to read them as an early and, perhaps, tentative statement of Leibniz's later more idealistic metaphysics. But if Leibniz is an idealist in those writings, he certainly isn't obvious about it; *never* in the CA does he *tell* Arnauld that all reality is mental or that everything that really exists is mind-like, doctrines that he will later explicitly embrace. I don't want to enter into a detailed comparison of the two apparently different metaphysical positions yet. A full comparison will turn out to be quite subtle; the metaphysical picture of the CA will, in the end, turn out to have more in common with the metaphysics of the *Monadology* than at first appears, as we shall see below in §IV. But it should be obvious that there is at least an *apparent* gap between the two positions.

But as different as the metaphysics of the CA and that of the *Monadology* appear to be, there is another metaphysical system to which the metaphysics of the CA bears an obvious affinity. What I have in mind is Leibniz's debt to the Aristotelian tradition. In an obvious way, Leibniz's position in the CA represents a conscious revival of the Scholastic version of the Aristotelian conception of substance. According to that conception, the substances that make up our world are composed of form and matter. For example, the Conimbrian Fathers wrote in their commentary on Aristotle's *Physics*:

> Natural things are not composed of matter [*materia*] alone, since if that were so, human beings, stones, and lions, being made of the same matter, would all have the same essence and definition. Therefore, in addition to matter, they have their own forms which differentiate them from one another.[43]

Or consider the account that Eustacius a Sancto Paulo gives in his *Summa philosophica*, a popular 17th century scholastic textbook:

> [A form is] a simple substantial actuality constituting, with matter, something that is a *per se* unity [*actus simplex substantialis unum per se cum materia constituens*].... Therefore, a form is a sort of substantial actuality, but, however, incomplete [*incompletus*], that is, an incomplete substance or, so to speak, a semi-substance, which, joined with matter constitutes one complete substance [*una integra substantia*]. And, it is allowed, this complete substance, depends in its composition on both parts, but that in its ground

and nature [*in sui ratione et natura*] it depends especially on its form.[44]

An obvious acknowledgment of Leibniz's debt to the Aristotelian conception of substance is his repeated use of the term "substantial form" to characterize the immaterial substance which produces unity in a corporeal substance.[45] Leibniz is fully aware of the Scholastic overtones of this term, and is perfectly willing to acknowledge and defend the revival of what his mechanist contemporaries would have seen as an anachronism. Leibniz writes in the DM:

> I know that I am advancing a great paradox in seeking to restore the old philosophy in some respects and to restore these almost-banished substantial forms. But perhaps I shall not be condemned so lightly when it is known that I have given much thought to the modern philosophy and...was for a long time convinced of the emptiness of these beings to which I am at last compelled to return in spite of myself and as by force. This is after I have myself carried out studies which convinced me that our moderns do not do enough justice to St. Thomas and other great men of his time and that the opinions of the Scholastic philosophers and theologians are much sounder than has been imagined, provided that they are used appropriately and in their place.[46]

But Leibniz's debt to the Aristotelian tradition also extends to his conception of matter as well. In an obvious way, just as the form of a corporeal substance is intended to correspond to the Aristotelian's notion of form, the body to which that form is united is intended to correspond to the Scholastic's matter: "for as our body is the matter, and the soul is the form of our substance, it is the same with other corporeal substances."[47] Of course, it isn't clear precisely how Leibniz understands "matter" here; we must in later sections explore that question and determine the extent to which Leibniz understands matter to be something genuinely distinct from minds or souls or forms. But whatever we conclude on that score, it is obvious that the metaphysics of the CA, the world of corporeal substances, is at least *prima facie* similar to the Scholastic-Aristotelian world. In recognition of the tradition to which Leibniz consciously and repeatedly appeals, I think that it is fair to call the metaphysics of the CA "Aristotelian" to distinguish it from the later, more (obviously?) idealistic monadology, though this is not to deny that there are many respects in which Leibniz departs from the Aristotelian position.

This preliminary account of the metaphysics of the DM and CA is somewhat surprising; we have found something that appears, at least at first glance, significantly different from the metaphysics more often associated with Leibniz's so-called mature writings. While, as I remarked earlier, I would like to put off a full account of the differences between the two apparently different metaphysics, it is necessary to address a general objection to the reading I have given. It is often said that Leibniz tailored his thought, or, at least, the way in which he expressed his thought to the intended audience of his writings. Can this explain what I have found in the CA? It is unlikely that Leibniz presented himself to Arnauld as an Aristotelian because he thought that Arnauld would be more sympathetic to him than if he presented his doctrine unadorned. While Arnauld was certainly familiar with the ins and outs of Scholastic metaphysics, he wasn't particularly sympathetic to that tradition; he was largely a Cartesian in matters philosophical, even if he didn't follow Descartes in all respects. Leibniz certainly knew that presenting his doctrine in Aristotelian language to a philosopher like Arnauld is hardly the best way to gain a sympathetic hearing. And, indeed, when Leibniz begins to use Scholastic terminology in the CA, Arnauld's reply is an impatient recitation of standard objections.[48] Now, if Leibniz thought that the Aristotelian language was merely an expository device, then this would have been the time to abandon it, and try some other approach. The fact that Leibniz persists in using such terminology and attempts to answer Arnauld's objections suggests that he thought that the Aristotelian framework of form, matter, and corporeal substance was the *right* way to express his doctrine. But, on the other hand, one might recall what Leibniz wrote a few years earlier in a letter to Herman Conring:

> Whenever I discuss matters with the Cartesians, certainly, I extol Aristotle where he deserves it and undertake a defense of the ancient philosophy, because I see that many Cartesians read their one master only, ignoring what is held in high esteem by others, and thus unwisely impose limits on their own ability.[49]

This indeed gives Leibniz a motivation for presenting his doctrine in Aristotelian language to a Cartesian like Arnauld. But, it is important to note, this motivation in no way undermines the claim that Leibniz saw certain important affinities between his doctrine and that of the

Scholastics. In fact, it is *because* he thinks that there is something importantly correct about Scholasticism that he is eager to defend it. It is still possible, of course, that this Aristotelian language is just an alternative way of expressing his own non-Aristotelian metaphysics; it is still possible that somehow underlying the world of Aristotelian corporeal substances is the familiar world of monads. This is, indeed possible, and will remain possible to the end; very little of what I say (or, more importantly, what *Leibniz* says) will discourage the interpreter who is determined to see the standard reading of Leibniz's metaphysics lurking behind the words he writes to Arnauld. But I don't think that this is the proper way to approach the text. We must remember that in the DM and CA, Leibniz uses Scholastic language, refers to Scholastic thinkers, and tells the reader that he is attempting to revive certain aspects of Scholastic metaphysics that, he claims, were unwisely rejected by his contemporaries. Nowhere in these writings is there any clear indication that he is doing *anything* else; any interpretation of the metaphysics of the CA in terms of the *Monadology* must be just that, an interpretation of one text in terms of others. Such interpretation is not always improper, and, in the end we may be driven to such a position in the case of the Aristotelian metaphysics that Leibniz appears to espouse in the CA and related writings. But I believe that our first obligation as commentators is to try to understand these texts on their own terms; when Leibniz associates his doctrine with that of the Scholastics, we should believe him, at least at first, and try to interpret his doctrines accordingly.[50] That is what I shall attempt.

This concludes my preliminary account of the metaphysics of the CA. There are, of course, many subtleties left to fill in, though. One concerns Leibniz's motivation. In emphasizing the aggregate argument that leads Leibniz to posit corporeal substances in the CA I don't mean to suggest that this is the *only* motivation for the position. There are at least two others suggested in the DM, CA, and other writings of the period, one approach relating to the broadly logical characterization of individual substance in DM8, and the other relating to the notion of activity and to Leibniz's program for physics, his dynamics.[51] These appraoches to corporeal substance are important, though, I think, more obscure than the relatively straightforward aggregate argument. However, I would like to put off considering them until later, when they will come up in the course of other discussions. More urgent is the project of fleshing out the rather bare-bones account I have given of Leibniz's world in the CA. In §II we shall deal in more detail with Leibniz's conception of corporeal substance and body, discussing matter, form, their union, and the different senses in which body and matter are taken to be phenomenal. In §III we shall discuss the fit between the notion of corporeal substance in the CA and the principal properties that Leibniz attributes to individual substances in the DM, particularly their indivisibility and immortality. And finally, in §IV we shall discuss the historical and philosophical relations between the Aristotelian metaphysics of the CA and the more apparently idealistic metaphysics of the monadological writings, before returning in the final sections to the questions about Leibniz's physics that started us off. I shall, in the end, argue that what I have called the Aristotelian picture provides us with an historically plausible and philosophically interesting way of looking at Leibniz's conception of the foundations of physics.

II. Form and Matter, Soul and Body

In the previous section, I outlined Leibniz's conception of the world as he relates it to Arnauld, a world packed at every level with corporeal substances, form and matter. In this section I would like to begin the task of unpacking that world a bit, dissecting its constituents and seeing how their parts fit together. We shall begin by examining Leibniz's conception of substance as made up of form and matter, united to produce an *unum per se*, a genuine entity.

Let us begin by introducing a distinction from the Scholastics that will be of some help in sorting out Leibniz's thought. The Scholastics drew a distinction between two conceptions of matter, primary (prime), and secondary. Eustacius, for example, writes:

> [Matter] is distinguished into primary and secondary. Primary [matter] is said to be that which, before all else, we conceive as entering into the composition of any natural thing, regarded as lacking all forms....Secondary [matter] is said to be that very primary [matter], not, however, bare, but endowed with physical actuality [i.e., forms].[52]

We shall have to look more closely at these concepts later, particularly that of primary matter. But the distinction is clear enough. Secondary matter is itself substance, the simple substance(s) out of which more complex things are made, the bronze or wood or plastic out of which a statue or house or pair of sunglasses is made; secondary matter is matter in the familiar sense in which we talk about the material out of which anything is made.[53] Secondary matter, thus, must include form, that which gives the material the nature it has. Primary matter, on the other hand, is the matter in the truest sense; it is what we get when we strip something of *all* of its forms (*per impossibile*).

This is a distinction that Leibniz finds helpful in clarifying his metaphysics in the CA. Leibniz writes:

> ...[E]xtended mass, considered without entelechies...is not a corporeal substance, but an entirely pure phenomenon, like the rainbow; therefore philosophers have recognized that it is form which gives determinate being to matter... But if one considers as the matter of the corporeal substance not mass without forms, but a second matter [*une matière seconde*] which is the multiplicity of substances of which the mass is that of the total body, it may be said that these substances are parts of this matter, just as those which enter into our body form a part of it, for as our body is the matter, and the soul is the form of our substance, it is the same with other corporeal substances.[54]

Though Leibniz is not entirely explicit here (he does not use the term primary matter), it is clear that the distinction he is drawing is the same one we saw in Eustacius, between matter as substance, the material from which more complex things are made, and matter in the strict sense as that which is left when we extract all of the forms from a substance.

One thing that this passage from the CA suggests is that while primary matter *may* be relevant to the composition of the corporeal substance (a question to which we shall later return), it is secondary matter which is most relevant. That is, it seems that the form of a corporeal substance unites not bare primary matter; rather, what it unites is secondary matter, a collection of substances, each of which has its own form and its own (secondary) matter. Thus, for example, when Leibniz talks of a corpse, an animate body detached from its soul, he writes:

> I admit that the body apart, without the soul, has only a unity of aggregation, but the reality remaining to it comes from its constituent parts which retain their substantial unity because of the living bodies which are included in them without number.[55]

And similarly, consider a passage we have seen in §I. Arnauld understood Leibniz's position as I have been reading it, seeing the corporeal substance as form uniting other lesser corporeal substances. This formed the basis of the following objection. Arnauld wrote:

> You admit these substantial forms only in animate bodies. Now, there is no animate body which is not organized [*organisé*], nor any organized body which is not many entities. So, far from your substantial forms preventing the bodies to which they are joined from being many entities, the bodies must be many entities in order to be joined to them.[56]

Leibniz's answer is instructive. Rather than denying that the body to which the soul is attached is made up of many entities, he is quite happy to recognize that it is, both in man and in other corporeal substances. Leibniz writes:

> I reply that assuming that there is a soul or entelechy in animals or other corporeal substances, one must argue from it on this point as we all argue from man, who is an entity endowed with genuine unity conferred on him by his soul, notwithstanding the fact that the mass of his body is divided into organs, vessels, humors, spirits, and that the parts are undoubtedly full of an infinite number of other corporeal substances endowed with their own entelechies.[57]

Corporeal substances are, then, best thought of as mind-like forms, uniting secondary matter, a collection of lesser substances, to form a genuine unity. But, one might ask, what is it that unites the form to the secondary matter? This is, of course, a problem that is not peculiar to Leibniz's metaphysics; it is a problem faced by the Scholastics of Leibniz's day,[58] and is, in a way, just the problem of what connects the mind to the body so familiar from Descartes and later Cartesians. An obvious candidate for an answer to this question is Leibniz's doctrine of pre-established harmony or, as it was called in the mid-1680's, the hypothesis of concomitance. That is, it is a certain harmony between the states of the immaterial substance that is the form, and the states of the corporeal substances that make up its body that seems to unite the form to its organic body in corporeal substances at every level.[59] Leibniz is not ***absolutely*** clear about this, and later he will deny that he ever intended pre-established harmony to account for unity in any strong sense.[60] But in the DM and CA, it certainly *looks* as if this is exactly what Leibniz has in mind.

This is clearest when Leibniz is discussing the human being. When

Leibniz first presents the hypothesis of concomitance in DM 33, it is presented as an "unexpected clarification of the *union* of the mind and body." Since the result of a union is a unity, and a genuine unity is, for Leibniz, a substance, this suggests that the hypothesis of concomitance is supposed to account for the fact that mind and body together constitute a substance. And consider a passage that Leibniz wrote to Arnauld in concluding an extended discussion of the hypothesis of concomitance. Leibniz wrote:

> The soul, however, is nevertheless the form of its body because it is an expression of the phenomena of all other bodies in accordance with the relationship to its own.[61]

The claim is that it is *because* of the special relation that the human soul bears to the human body, the fact that the soul expresses the universe by expressing its body, the relation that constitutes harmony, that the soul is the form that belongs to the body to which it is attached. The suggestion is that it is the harmony which attaches the soul to the body in such a way as to make it the form of the body, and to make it *a* form with respect to the substance which is the human being.

Leibniz says nothing *explicit* to suggest that harmony is what unites form to matter in the more general case of a non-human corporeal substance. But if, indeed, harmony is what is supposed to unite the human soul to the human body in such a way as to form a genuine corporeal substance, then there is every reason to think that Leibniz would extend the account to the more general case, at least in the CA. Leibniz emphasizes in a number of places that his hypothesis of concomitance is not a *special* hypothesis about human minds and human bodies, but a "consequence of the concept of an individual substance."[62] Furthermore, as I have noted earlier, Leibniz suggests a number of times to Arnauld that corporeal substances, both their constituents (form and matter) and their unity can be understood on analogy with the corresponding aspects of human beings. Thus form is somewhat like the human mind, matter like the body. And, to extend the analogy, one would suppose that their unity consists in a harmony between what goes on in the one and in the other. This, at any rate, is what the texts suggest. So far we have been playing one side of the distinction between primary and secondary matter. I have argued that the body of a corporeal substance is to be thought of as a *secondary* matter, as an aggregate of lesser corporeal substances, united by a form to produce a corporeal substance, and have suggested that what may unite them is just pre-established harmony or concomitance, a particular relation between the immaterial substance that consitutes the form of the corporeal substance, and the corporeal substances that together constitute the body. But what of *primary* matter. How does *that* enter into Leibniz's metaphysics?

The notion of primary matter in late Scholastic philosophy is complex, and it is not possible to give anything like a complete account of that notion here.[63] Unlike secondary matter, primary matter is not itself a substance or a collection of substances, but, together with form, a constituent of substance, and it is in contrast with form that it must be understood. Among the Scholastics of the 17th Century, three contrasts seemed to be of importance. First of all, primary matter is taken to be purely potential (*potentia*, power) and purely passive, "having no effective force."[64] Matter in this sense is taken to be actualized by form, or pure activity or actuality. Secondly, it is, in contrast with the form, which, as we have seen, provides *unity* to the substance, that from which the quantity or extension follows. "Matter alone has extended quantity *per se*, and through its pure substantial being," Suarez wrote, "and everything else which is in matter [i.e., form] is quantitative *per accidens* and lacks true quantity."[65] And finally, matter is contrasted with form as subject to property. Primary matter understood in this sense is the ultimate subject to which forms attach. Eustacius, for example, argues that since a physical composite (secondary matter) cannot progress, substance in substance, to infinity, there must be some bottom-most substances. These bottom-most substances are forms attached to the primary matter, the "*primum subjectum & materia prima*" as he calls it.[66]

These characterizations of primary matter are, to be sure, obscure and in need of further elucidation. But they give us enough of a grasp on the notion to raise what is, to my mind, a crucial question for Leibniz interpretation in the period of the DM and CA. As we have seen, Leibniz clearly recognizes something that he identifies with the Scholastics' form and with the Scholastics' secondary matter, unities that can unite multiplicities of entities, and multiplicities that can be united by unities. But does Leibniz recognize anything like primary matter?

The standard answer to this question is no. It is acknowledged by some commentators (e.g., Russell) that Leibniz recognizes something that, from time to time, he is prepared to *call* primary matter. On this reading, primary matter is, for Leibniz, just the monad or immaterial substance insofar as it is being acted upon rather than acting.[67] Given Leibniz's account of activity and passivity in terms of clear and confused perceptions, this is to say that "substances have metaphysical matter or passive power insofar as they express something confusedly; active, insofar as they express it distinctly," as Leibniz wrote in an important but, unfortunately, undated piece.[68] But, it is claimed, primary matter for Leibniz is not material in any proper sense; it is not a first subject, nor is it connected with extension in any way, nor does it lead Leibniz to recognize anything in nature over and above the immaterial substances he later calls monads. On this reading, the corporeal substance of the CA may be understood as a form attached to a body, but the body itself must be interpreted as an aggregate of forms. Corporeal substances, then, are immaterial substances (form) dominating other immaterial substances (the forms of the substances making up the body or secondary matter) which dominate further immaterial substances, and so on down. And so, the claim is, Leibniz has no need for anything but the immaterial substances he calls souls, forms, or entelechies; no need, on this view, for anything, like primary matter, that is defined in contrast to forms.

There is, indeed, support, both direct and indirect, for the claim that this is what Leibniz had in mind in the CA. Leibniz says, for example:

> Extended mass, considered without entelechies...is not a corporeal substance, but an entirely pure phenomenon like the rainbow; therefore philosophers have recognized that it is form which gives determinate being to matter...[69]

Similarly, Leibniz writes to Arnauld:

> ...the matter considered as the mass itself is only a pure phenomenon or a well-founded appearance [*apparence bien fondée*], as also are space and time.[70]

This *appears* to be as direct a rejection of primary matter as one could want. And, less directly, there seems no room in Leibniz's metaphysics for primary matter, in at least one of its important senses. Since Leibniz recognizes substances within substances *ad infinitum*, there seems to be no room for a first subject, primary matter, in his system. And without primary matter, all that seems left is form, soul, entelechy.

These arguments and the interpretation that they lead one to are plausible, to be sure. But I don't think that they are decisive. While there is little question but that Leibniz of the *Monadology* adopted a kind of idealism, there is, I think, reason for thinking that Leibniz *may* have recognized the primary matter of the Scholastics as an important part of his metaphysics in the period of the DM and CA. Leibniz, of course, needs no first subject, no ultimate ground to which forms can attach themselves. But, as I shall argue, Leibniz does feel the need to recognize in bodies something over and above the immaterial substances he calls forms, something from which extension can arise, something which, in later sections, we shall connect with passivity. And, I shall argue, the sense in which this primary matter is unreal and phenomenal is consistent with its being a genuine component of the corporeal substance, and its being, in an important sense, genuinely distinct from form.

Let me begin by pointing out, once again that *nowhere* in the DM and CA does Leibniz *ever* say that the ultimate constituents of reality are souls or soul-like, something that he will later emphasize quite explicitly and at great length. The innocent eye reading those documents could hardly accuse Leibniz of such extravagances; it is, I think, only in the light of our knowledge of other texts Leibniz wrote, most notably the *Monadology*, that we are tempted to read this idealism into those earlier texts. Unfortunately, though, Leibniz doesn't argue the contrary in the DM or CA either; while there is no explicit suggestion of his later idealism, of the claim that everything is mental, there is precious little discussion of the non-mental, material component of reality. However, the question does come up, if somewhat obliquely, in a closely related text. In March 1690, Leibniz wrote an important letter to Michelangelo Fardella.[71] Leibniz seems to have met Fardella in Italy, where he traveled from 1687 to 1690, and seems to have discussed with Fardella many of the same questions he discussed in his letters to Arnauld.[72] The letter in question consists of a brief summary of Leibniz's position, together with a statement of some objections and questions Fardella raised and Leibniz's answers. Leibniz's answers are not always clear and consistent, particularly when dealing with his theory of body.[73] But in that document, Leibniz does suggest an intriguing analogy between geometrical points on the one hand, and souls on the other that strongly hints at the necessity for something non-mental in body, something closely connected with the extension bodies are observed to have.

But in order to appreciate Leibniz's analogy here, we must first say something about Leibniz's conception of points and lines. A full consideration of this question would take us into Leibniz's solution to the problem of the continuum, a topic far beyond the scope of this paper. Put briefly, though, mathematical points are not geometrical *objects*, strictly speaking. "Points are not parts [of a line] but limits [*termini*]," Leibniz writes, or "extremities" as he calls them elsewhere.[74] Leibniz's idea here is that points are what mark the boundaries of line segments and, as such, are "modifications" or "modalities" as he sometimes calls them[75] rather than geometrical objects, like lines. Consequently, Leibniz argues, extension cannot arise from points alone: "even an infinity of points gathered into one will not make extension."[76] The situation is, in fact, quite the reverse; it is from extension that points arise, as the termini of finite line segments, and not *vice versa*. Thus Leibniz argues that while there is no finite bit of geometrical extension that doesn't include an infinite number of points, the line is not composed of points alone:

> Extension arises from situation, but it adds continuity to situation. Points have situation, but they neither have nor compose continuity, nor can they stand by themselves.[77]

So, there are points everywhere in extension insofar as a line is arbitrarily divisible into smaller segments *ad infinitum*. But geometrical extension must include something more than points, something in itself extended from which these points arise.[78]

Now we can turn to what Leibniz writes to Fardella:

> What points are in the imaginary resolution [of the line], souls are in the true [resolution].[79]

Given what we have seen of Leibniz's conception of the soul or form as that which unifies an organism, a corporeal substance, and given Leibniz's conception of the point, this analogy has an obvious meaning and important consequences. Insofar as points are extremities of line segments, they can be said to individuate line segments, the parts of which longer lines are made up, by fixing their termini; it is in this way that points serve to resolve a line into smaller parts in geometry, an "imaginary resolution" insofar as we are considering the line abstracted from any and all real extended things. Leibniz's claim is that in real extended things, bodies, souls perform something of the same function that points perform in abstract "imaginary" extension. So, the claim seems to be, what marks out the *real* divisions in bodies are souls, that which unifies corporeal substances and makes genuine individuals of them. So far this is not very surprising; we have seen already and at some length the connections Leibniz sees between souls, individuality, and the makeup of bodily composites. But Leibniz draws a further conclusion from this analogy. While he recognizes important differences between points in geometry, and souls or forms in real things, the analogy extends at least this much farther:

> And therefore, there are substances everywhere in matter, as points are in a line. And just as there is no portion of a line in which there are not an infinite number of points, there is no portion of matter which does not contain [an] infinite [number of] substances. But just as a point is not a part of a line, but a line in which there is a point [is such a part], so also a soul is not a part of matter, but a body in which there is a soul [is a part of matter].[80]

And thus Leibniz opens his answer to Fardella's objections by claiming:

> I don't say that body is composed of souls [*anima*], or that body is constituted by an aggregate of souls, but of substances.[81]

The claim seems to be that just as points in geometry cannot make up geometrical extension, souls, necessary as they are in the *individuation* of corporeal substances, the real parts or constituents of bodies, are not sufficient to make up real extension by themselves.[82] The suggestion is that just as there must be something more to geometrical extension than points, there must be something more to the real extended thing, the body, than the souls that individuate its constituents, something intimately connected with its real extension, presumably. This argument, unlike the Scholastics' "first subject" conception of primary matter, doesn't tell one *where* in the corporeal substance this non-mental principle of extension is to go in the infinite hierarchy of corporeal substances contained in any body. But it *does* claim to establish that it must be *somewhere*, that souls by themselves are not sufficient to constitute corporeal substances.[83]

This bit of reasoning that Leibniz offered Fardella is not found in the CA. But there are passages that hint at the existence of something connected with bodily extension, something over and above the soul or form that is not unlike the Scholastic's primary matter. Consider, for example, this passage from the CA where Leibniz is apparently endorsing a sense of the term "matter" that he explicitly distinguishes from secondary matter:

> But if one were to understand by the same term 'matter' something that is always essential to the same substance, one might in the sense of certain Scholastics understand thereby the primitive passive power of a substance, and in this sense matter would not be extended or divisible, although it would be the principle of divisibility or of that which amounts to it in the substance.[84]

This sentence is obscure in many ways, and we shall have to return to it below in §V when we discuss passive force and its connection to extension. But in its contrast with secondary matter, in the connections drawn between matter, extension, passivity, and potentiality (*potentia* is, remember, ambiguous between power and potentiality), Leibniz seems here to be acknowledging that there is something in the world over and above bare soul or form, something that is a recognizable version of the primary matter of the Scholastics.

I have tried to establish that for Leibniz in the period of the DM and the CA, forms are not enough to constitute the world of bodies, and that Leibniz recognizes the necessity for something like the primary matter of the Scholastics, in addition to form. But, one might well ask, what about the passages I cited earlier in which Leibniz seems to *reject* primary matter, matter without forms? Such passages are significant, to be sure. But their precise meaning and import depends on Leibniz's conception of reality, what is real and what is, as he puts it, merely phenomenal.

Leibniz uses the term "phenomenal" in two distinctly different senses in the CA.[85] In one sense of the term, phenomena are equivalent to illusions, the perceptions of a conscious mind that don't correspond to anything in the world external to the perceiver. Thus Leibniz writes to Arnauld:

> You object, Sir, that it may be of the essence of body to be devoid of true unity; but it will then be of the essence of body to be a phenomenon, lacking all reality as would a coherent dream, for phenomena themselves like the rainbow or a heap of stones would be wholly imaginary if they were not composed of entities possessing true unity.[86]

It is obvious how dreams are phenomenal in this sense. Leibniz's point is that if there were no unities, no substances in bodies, then bodies would be phenomenal in this extreme sense, no more real than dreams.

But there is another sense in which things can be phenomenal. This is the sense in which the bodies of our world, made up as they are of corporeal substances are less real than the corporeal substances that make them up. Put briefly, bodies are phenomenal in this sense insofar as they are unified by *us*, not by nature; they are collections of genuine unities that are united by an act of mind. Thus Leibniz writes:

> Only indivisible substances and their different states are absolutely real. This is what Parmenides and Plato and other Ancients have indeed recognized. Besides, I grant that the name of 'one' can be given to an assembly of inanimate bodies, although no substantial form links them together, just as I can say: there is *one* rainbow, there is *one* flock; but it is a phenomenal or mental unity [*unité de phenomene ou de pensée*], which is not enough for the reality in phenomena.[87]

Or, Leibniz wrote at much greater length:

> It can therefore be said of these composite bodies and similar things what Democritus said very well about them, 'they exist by opinion, by convention [*lege*] -- *nomo*.' And Plato holds the same view about what is purely material. Our mind notices or conceives of certain genuine substances which have various modes; these modes embrace relations with other substances, from which the mind takes the opportunity to link them together in thought and to enter into the account one name for all these things together, which makes for convenience in reasoning. But we must not let oneself be deceived and make of them so many substances or truly real entities; that is only for those who stop at appearances, or those who make realities out of all the abstractions of the mind....[88]

A body considered as an aggregate of substances is phenomenal in this sense insofar as we impose the unity on it; we make it one, when all there really are are relations among different substances. If we think of a rainbow not as the array of color, but as the collection of water droplets that causes the observed rainbow,[89] then it too may be considered as phenomenal in this sense, insofar as it is the mind that unites the droplets to form *one* thing, *one* rainbow.

In developing this second sense of the term phenomenal, Leibniz emphasizes its application to bodies as aggregates of substances, united by the mind. But the term has a more general signification, I think. Leibniz writes, shortly after the longer passage quoted:

> I maintain that one cannot find a better way of restoring the prestige of philosophy and transforming it into something precise than by distinguishing the only substances or complete entities, endowed with genuine unity...; all the rest is merely phenomena, abstractions or relations.[90]

It is obvious how bodies fit this account; it is through the perception of the relations between genuine substances that we are led to posit an entity. But Leibniz also means this account to apply to other notions. In a passage found earlier on in the same paragraph from which the last quotation was taken, Leibniz contrasts those (like himself) who take only substances to be real with "those who make realities out of all of the abstractions of the mind, and who conceive of number, time, place, movement, shape...as so many separate entities."[91] What Leibniz seems to have in mind here is that the modes of space and time are phenomenal insofar as they arise from abstracting a mode or property from the thing of which it is a mode or property and treating it as if it were a real thing, as when we separate number from things numbered or shape from things that have shape and treat them as genuine entities, the number 'four' or the shape 'square.' One can say the same thing about space and time themselves, insofar as they are regarded as real things. What this has in common with the phenomenality of body is that in both cases something is posited as an entity, as a genuine *thing* by virtue of an act of mind, whereas in reality, the "thing" in question is just mode or a collection of things. This is, I think, the best way of understanding this sense of the phenomenal, the sense in which something can be phenomenal without being entirely illusory on that account.[92]

What, then, are we to make of the apparent denials of primary matter? One might read Leibniz as claiming that primary matter is illusory, phenomenal in the first sense I outlined earlier. But the texts suggest something different, I think. In one of the texts cited earlier, Leibniz claims that (primary) matter is a "pure phenomenon or a well-founded appearance."[93] In this respect, Leibniz says, matter is in the same boat as space and time.[94] But space and time are most naturally construed as phenomenal in the second sense I outlined earlier. Space and time seem to be phenomenal not in the sense of being wholly imaginary; they are phenomenal insofar as they exist as entities, rather than as modes or relations, by virtue of an act of mind. The explicit linking of primary matter with space and time suggests that primary matter might be phenomenal in a parallel sense.

It is not difficult to see how one can construe primary matter as phenomenal in the sense of a reified abstraction, like space and time. On a familiar Scholastic doctrine, primary matter cannot exist by itself and without form; primary matter, on this view, is a constituent of substance, but not, properly speaking, *a thing*, insofar as it lacks the capacity for independent existence.[95] This doctrine is what Leibniz seems to have in mind in a passage quoted earlier in connection with the phenomenality of body, in which Leibniz makes reference to the doctrine of "the philosophers" in accordance with which "it is form that gives determinate being to matter."[96] Translated into more straightforwardly Leibnizian terms, matter cannot exist without form because form is that which gives unity, and nothing without unity can *really* exist.[97] Understood in this way, primary matter is an abstraction from substance taken in its entirety, and considered as a *something*, it is phenomenal. Though it is not completely mind-dependent and imaginary, like a dream, for example, it is only an act of mind that can separate it from the substance of which it is a constituent and make it into a genuine entity.

And with this, it is now possible to state with some precision the sense in which Leibniz may have recognized primary matter as something over and above form or soul. Primary matter can never exist without form or mind; consequently, primary matter cannot be characterized as a something that can be added to form to produce a corporeal substance. But one can draw a real distinction between a form or soul with matter, and that same form or soul without; the former is a corporeal substance, the latter an immaterial substance.[98] Insofar as there is a real difference between the two kinds of substances, one can say that Leibniz recognizes the existence of primary matter. Primary matter is not a real *something* for Leibniz any more than it was for most of his Scholastic forebearers, to be sure. But it must in some sense exist as that which distinguishes a form in its extended and embodied state from that same form in its non-extended and disembodied state; it is that which corporealizes an incorporeal substance. I should remark that the evidence that Leibniz held such a position in the CA is by no means decisive, and I present the claim with some hesitation. But given Leibniz's clear identification of his position with that of Scholastic metaphysics, and given the absence of any clear statement of the later more idealistic position on matter, the reading is, at very least, a plausible conjecture.

III. Corporeal Substance as Individual Substance

I have been arguing that among the individual substances that Leibniz recognized in the DM and CA are corporeal substances, unities of form and matter, soul and body, that Leibniz understands on analogy with organisms. Now, in the DM Leibniz is quite explicit about what properties individual substances, as he understands them, are to have. Individual substances admit of a complete individual concept and contain marks and traces of everything that has happened, is happening, or will happen to them. They all differ from one another, they are all immortal, indivisible, and all mirror the entire world in which they find themselves. And finally, all of the states of an individual substance proceed from within the substance itself, and no individual substance exerts any causal influence over any other, properly speaking.[99] These claims have been examined at length in the context of the assumption that the individual substances of the DM are no different than the monads of Leibniz's later years. But if I am right about the place of corporeal substances in Leibniz's thought in this period, then such an account will not do: we must determine how these doctrines can apply to corporeal substances, individual substances that have a touch of the material. Some of Leibniz's claims are no more puzzling for corporeal substances than they are for monads, for example, the claim that no two individual substances can share all of their properties. Some of Leibniz's claims are, if anything, *more* plausible for corporeal substances than for monads. Leibniz himself often illustrates the non-communication claim by appeal to the phenomenon of elastic collision in which he claims "bodies strictly speaking are not pushed by others when an impact occurs, but by their own elasticity."[100] Similarly, Leibniz often appeals to the plenum of bodies in a physical world to argue for his mirroring claim:

> I believe that M. Descartes himself would have agreed with this, for he would undoubtedly grant that because of the continuity and divisibility of all matter the smallest movement extends its effect over neighboring bodies and consequently from neighbor to neighbor *ad infinitum*, but proportionately decreased; thus our body must be affected in a way by the

> changes of all the others.[101]

These properties, then, are relatively unproblematic in the domain of corporeal substances. But not so for some of the other properties of individual substance. In this section we must explore how Leibniz saw corporeal substances as indivisible and immortal, and how the doctrine of complete individual concepts and the closely related marks and traces claim apply to the quasi-Aristotelian organisms that, I have been arguing, make up Leibniz's world at this time.

Let me begin this enterprise by outlining what Leibniz says concerning the indivisibility and indestructibility of corporeal substances. Arnauld, naturally enough, finds the indivisibility claim somewhat puzzling. His first objection runs as follows:

> Is it the substantial form of a marble tile that makes it one? If that is the case, what becomes of this substantial form when it [i.e., the tile] ceases to be one because it has been broken into two?[102]

Leibniz answers:

> I believe that a marble tile is perhaps only the same as a heap of stones and thus cannot be considered a single substance but a collection of many.[103]

Arnauld's mistake is in thinking that Leibniz intended inanimate objects, like marble tiles, to be genuine substances, with their own substantial forms.[104] If this were Leibniz's position, then it would indeed be puzzling to consider corporeal substances as indivisible, since a single marble tile can, indeed, be split into two pieces, each of which is, it seems, a marble tile in its own right. But Leibniz's answer is straightforward: the marble tile lacks a single unifying form, and thus is not a corporeal substance, but an aggregate of corporeal substances. Consequently, Leibniz sees no problem here; the indivisibility claim only applied to the rudimentary organisms that Leibniz considers genuine substances. More relevant to Leibniz's position is a second question that Arnauld raises:

> For what reply can one make about those worms which are cut into two, each part of which moves as before?[105]

Leibniz answers:

> As regards an insect which one cuts in two, the two parts do not necessarily have to remain animate, although a certain movement remains in them. At least the soul of the whole

> animal will remain only in one part....[106]

Leibniz's position seems to be that when we are dealing with a genuine corporeal substance, like he supposes Arnauld's worm to be, then it *must be* indivisible; one cannot split one living thing to make two living things, Leibniz claims. When the body of a living thing is split, its soul, that which makes it the corporeal substance it is, must remain in one half or the other. Thus, in the case of Arnauld's worm, at most one half of the worm divided can remain animate and substantial; the motion that remains in the other half cannot be the motion of an animate creature.[107] Leibniz is in no way intending to deny that one can divide the body of a corporeal substance into smaller parts; one can take out a person's appendix, cut a flower off of a rose bush, or split a worm into two wriggling parts. But dividing the body of a corporeal substance is altogether different from dividing the substance itself; hard as one may try, Leibniz insists that the knife cannot make two living worms from one, two *substances* out of one. That is, corporeal substances are indivisible in the sense that one cannot take a corporeal substance and split it into two parts, each of which is equally well a corporeal substance, soul or form unifying a body.

And just as corporeal substances are indivisible, Leibniz claims that they are indestructible as well. This claim, too, is somewhat paradoxical. Corporeal substances are all, in a very general sense, living things, and Leibniz certainly cannot deny that living things die. But, Leibniz claims, the death of an organism is not the destruction of a corporeal substance, but only its transformation. Thus, when Arnauld asked what becomes of worms and their souls when burned in a fire, Leibniz replies:

> Those who conceive that there is almost an infinite number of little animals in the smallest drop of water, as the experiments of M. Leeuwenhoeck have made known, and who do not find it strange that matter is everywhere full of animate substances, will not find it strange either that there is something animate even in ashes and that fire can transform an animal and reduce it in size instead of totally destroying it.... [Such organisms are] little organic bodies, wrapped up as they are because of a sort of contraction from a larger body which has undergone corruption....[108]

Thus, "corruption or death is nothing other than the diminution and envelopment of an animal which nevertheless goes on surviving and remaining alive and organic."[109] Corporeal substances, thus, do not perish in Leibniz's world, but only grow or shrink.[110]

This, in brief, is Leibniz's position on indivisibility and indestructibility of corporeal substances. But it remains to see the philosophical basis on which it rests, a number of intertwined positions on souls and forms, the relation between souls and forms and their bodies, and the individuation of corporeal substances.

The first and most important of these positions is the claim that forms and souls are indivisible and naturally indestructible, a claim that he repeats a number of times in the CA.[111] Leibniz never attempts to produce an argument for that conclusion, though, perhaps because it is a position that he considers generally accepted by a wide variety of philosophers, ancient, modern, and medieval.[112] Secondly, Leibniz claims that while souls and forms taken apart from any bodies are genuine substances, immaterial substances, and, as such, can presumably exist without being attached to any body in nature, in *this* possible world, every soul is the soul of some body: "there is naturally no soul without an animate body"[113] Leibniz writes. Leibniz's motivation for this doctrine is unclear. It is possible that it derives from a simple application of the principle of perfection. The world Leibniz outlines in the CA is a tightly organized world, a world of corporeal substances nested in one another to infinity. A bare soul, existing in this order but detached from a body would seem to detract from the orderliness of the world. Another possible motivation for this doctrine, one found explicitly in somewhat later writings, derives from the role of the form as active principle and the body or its primary matter as passive. On this account, soul or form cannot exist in nature without body because the active is incomplete without the passive, or because only God can be completely active and non-passive.[114] Given the lack of attention to notions of activity and passivity in the CA, and given the attention Leibniz gives the principle of perfection there, this motivation seems less likely than the former, though. A third doctrine relevant here concerns the individuation of corporeal substances. While Leibniz is not explicit, he seems to take for granted throughout the discussion of corporeal substances that they are individuated by their forms. That is, a corporeal substance or organism can change its body as much as nature will allow, but as long as it has the form or soul it has, it is the same corporeal substance or organism. This principle of individuation seems to be an extension of what might be considered a simple-minded Cartesian principle of individuation for persons (same mind, same person) to the wider domain of corporeal substances, soul-like entities united to bodies.[115]

Together these three doctrines entail the indivisibility and indestructibility of corporeal substances. Since forms are indivisible, so must be the corporeal substances that they serve to individuate. In fact, it is *because* corporeal substances as substances must be indivisible that they must have forms, Leibniz sometimes argues:

> ...I think I have shown that every substance is indivisible and that consequently every corporeal substance must have a soul or at least an entelechy which is analogous to the soul, since otherwise bodies would be no more than phenomena.[116]

The indivisibility of corporeal substances thus derives from the indivisibility of their forms. Similarly, the indestructibility of corporeal substances derives from the indestructibility of the forms attached to their bodies. Since forms are indestructible, since they are always attached to some body, and since they are what identifies the corporeal substance as the substance it is, it follows that corporeal substances must also be indestructible.[117]

This conception of the indivisibility and immortality of the corporeal substance provides an interesting point of view from which to regard what we said earlier about form and its role in unifying a complex of lesser substances and producing a genuine individual, a genuine corporeal substance. The mark of individuality is the distinction that one can draw between one individual and another; individuality in the strongest sense is possible only when we can distinguish an individual from others at the same time and at different times. Form can be said to unify the corporeal substance by producing an individual in precisely that sense. The form fixes the corporeal substance as an individual persisting through time by providing a constant reference point: *this* corporeal substance is *this* form, together with whatever body (matter) it may now have. Form also gives an unambiguous criterion for determining whether or not two hunks of matter are part of the same corporeal substance at a given time: they are if and only if they are appropriately related to the same form. In this way the form allows the corporeal substance to be reidentified over time, and allows us to identify its parts or constituents, and distinguish them from unrelated hunks of body; in this way the form produces one entity from many.

And with this we can see more clearly how Leibniz's conception of corporeal substance is linked to the complete individual concepts (CIC), the marks and the traces that play such a prominent role in Leibniz's account of the individual substance in the DM and elsewhere. In DM8, Leibniz introduces the claim that every individual substance has a CIC, and the internal marks and traces by virtue of which that CIC applies to the substance that it is true of. After drawing out some consequences of this claim in DM9, Leibniz makes an apparent reference to it in DM10, linking the conception of substance developed in DM8 to the Scholastic doctrine of substantial form. Leibniz writes:

> Not only the ancients but also many able men given to deep meditation...seem to have had some knowledge of what we just said; this is why they have introduced and maintained the substantial forms which are so widely discredited today. But they are not so far from the truth, or so ridiculous, as our modern philosophers commonly imagine.[118]

The connection between the doctrine of substance announced in DM8 and the doctrine of substantial forms mentioned in DM10 is obscure. But it is somewhat illuminated in a passage that Leibniz wrote to Arnauld:

> Substantial unity requires a complete, indivisible and naturally indestructible entity, since its concept embraces everything that is to happen to it, which cannot be found in shape or in motion (both of which embrace something imaginary, as I could prove), but in a soul or substantial form after the example of what one calls Self.[119]

While this passage is not without its own obscurities, what it suggests is this. According to DM8, every individual substance must have a CIC, a concept that includes everything that was true, is true, or will be true of that substance, as well as marks and traces of all that was true, is true, or will be true of the substance in question, i.e. something *in* the substance by virtue of which the CIC is true of that substance. Suppose that it is *now* (at t1) true of substance S that S will at some future time (at t_2) have property P. Leibniz *seems* to be reasoning that if it is true at t_1 that S will have P at t_2, then there must be something *at* t_1 which *will be* P at t_2, and something at t_2 which *is* P, and these two somethings must be the same thing. If we are *now* to have a truth about something in the future, then there must be some one thing to which both the present and future facts attach; an enduring truth requires an entity that endures, for Leibniz. A similar argument can be given from the present truth of facts about the past. Since a CIC includes facts about the future (and past) states of a substance, there must be something that persists from the past, to the present, and into the future, something that is present whenever the substance is, something to which the past, present, and future properties can attach themselves. This something makes the substance the individual it is, and, since it is that which persists in a substance, it is that which serves as a permanent subject for the concepts that make up the CIC, and that in which we find the actual marks and traces by virtue of which the CIC applies. This much is true about all individual substances. Substantial forms enter when we consider how the conclusion applies to bodies. If bodies are to have something substantial, then they must contain something that persists, that allows reidentification of the corporeal substance from one time to the next, that can serve as a subject for the CIC. This something is, of course, the form or soul that unites the corporeal substance, and creates a genuine persisting individual; the matter in the body, fluctuating and changing from moment to moment just will not do, Leibniz thinks. And with this we have a connection between the doctrine of substance in the DM, and the account of form and corporeal substance so prominent in the CA. Since the form or soul is that which persists in a corporeal substance, the form or soul must be that to which the CIC attaches; it must, thus, *also* be that in which we find the marks and traces by virtue of which the CIC applies to the substance. It is, Leibniz quite clearly says, in the *soul* of Alexander, a representative corporeal substance, that we find "traces of all that has happened to him and marks of all that will happen to him and even traces of all that happens in the universe...."[120] And with this we have a second argument for why there must be something mind-like in bodies. If bodies are to contain something substantial, then they must, by DM8, contain something capable of a CIC; if they contain something capable of a CIC, then they must contain something that persists and enables us to reidentify the corporeal substance from one time to the next, a "complete, indivisible, and naturally indestructible entity," not matter, but "a soul or substantial form after the example of what one calls *Self*."[121] The argument may not be without its gaps and logical lapses; it is not an argument I would want to defend. But it does explain why Leibniz saw an intimate connection between the so-called logical doctrine of substance given in DM8 and the Aristotelian doctrine of substantial forms that plays such a prominent role in his account of corporeal substance.

IV. Some Historical Considerations: Aristotelianism and Idealism

In the previous sections I have been sketching Leibniz's world mainly as it appears in the DM and CA, the two important documents from the mid and late 1680's. In this section I would like to raise some questions about Leibniz's philosophical development, about the way that the position I have been sketching fits into the larger context of Leibniz's work. I shall begin by trying to differentiate the position I have sketched earlier from what is commonly accepted as Leibnizian metaphysics, the metaphysics sketched in the *Monadology* and related writings. I shall then investigate the extent to which we can set boundaries on Leibniz's Aristotelianism, and specify when he definitely held it and when he gave it up.

The world I have been sketching, a world of both immaterial and corporeal substances, a world of form, matter, and quasi-Aristotelian substances is, at first glance, very different from the more familiar Leibnizian world of the later writings and the standard commentaries. But the two worlds have much in common, and it is not easy to put one's finger on how exactly they differ. The world of the CA and the world of the *Monadology*, to choose a work representative of Leibniz's later thought, both contain immaterial substances. And while commentators have not emphasized this aspect of Leibniz's later thought, the world of the *Monadology* is as full of organisms as is the world of the CA: once Leibniz fills his world with life, it *remains* full. In the *Monadology*, every monad has an organic body, and the body associated with every monad is, itself, a collection of organisms, further monads and their bodies.[122] Thus, Leibniz can write in the *Monadology* a passage very reminiscent of what he wrote to Arnauld:

> It is clear from this that there is a world of creatures, living beings, animals, entelechies, souls, in the smallest particle of matter. Each part of matter can be thought of as a garden full of plants or as a pond full of fish. But each branch of the plant, each member of the animal, each drop of its humors, is also such a garden or such a pond.[123]

This much of the picture is clearly common to both worlds.

But there is also a fundamental difference, I think, a difference which is signaled in the opening sections of the *Monadology*. Leibniz writes:

> The *monad* which we are to discuss here is nothing but a simple substance which enters into compounds. *Simple* means without parts. There must be simple substances, since there are compounds, for the compounded is but a collection or *aggregate* of simples. But where there are no parts it is impossible to have either extension, or figure, or divisibility. The monads are the true atoms of nature; in a word, they are the elements of things.[124]

And, a few sections later Leibniz writes:

> This is the only thing -- namely, perceptions and their changes [i.e., appetitions] -- that can be found in simple substance. It is in this alone that the *internal actions* of simple substances consist. All simple substances might be given the name of entelechies.... There is in them a certain sufficiency which makes them the sources of their internal actions and so to speak, incorporeal automata. If we wish to designate by soul everything which has perceptions and appetites..., all simple substances or created monads could be called souls.[125]

To put it simply, the world of the *Monadology* is made up of monads, simple substances that are uncompounded of other substances, non-extended immaterial substances that are basically mental in nature. And, strictly speaking, *that is all that there is*. As Leibniz wrote to DeVolder in 1704: "Indeed, considering the matter carefully, it may be said that there is nothing in the world except simple substances and, in them, perception and appetite."[126] The world of the *Monadology* is, thus, a world of souls, and the metaphysics a variety of idealism.

Leibniz's later metaphysics involves two significant differences with respect to the metaphysics of the CA. In the CA, I have argued, Leibniz recognizes two distinct and genuinely different varieties of substance; there are minds, souls, forms, entelechies, immaterial substances understood on analogy with the immaterial "self" we each find within, and corporeal substances, forms or souls united to bodies, and understood on the model of living things. One can say, in fact, that in the CA, at least, Leibniz's *main* interest is in the latter, the animate and corporeal unities that ground the reality of the world of bodies; it is the corporeal substances, and not their forms or souls, that may be considered "the true atoms of nature," to use Leibniz's later phrase. The incorporeal substances seem *mainly* of interest to Leibniz in the CA insofar as they go to create such animate unities. The position is significantly different in Leibniz's later writings. There the clear emphasis is on the simple substances, the monads. There are complex collections of monads which constitute living things, to be sure. But Leibniz is no longer *certain* that they are genuine substances, and it no longer seems to be a central part of his doctrine that they must be substances and that such corporeal substances must ground the reality of the bodies of everyday experience. Sometimes Leibniz does call such organized collections of monads substances, compound substances as opposed to the simple substances that make them up. In the *Principles of Nature and Grace*, for example, Leibniz writes:

> And each separate simple substance or monad which forms the center of a compound substance (such as an animal, for example)...is surrounded by a mass [*masse*] composed of an infinity of other monads, which constitute the body belonging to this central monad....[127]

But such composite entities are clearly problematic in a way in which they were not in the confident letters to Arnauld.[128] In the *Monadology*, for example, the monad and its organic body are not called a *substance* but simply a "natural automaton." And, of course, the very possibility of a composite substance, a substance composed out of monads, is a central worry of the troublesome correspondence with DesBosses. Furthermore, even when Leibniz grants that there are composite substances in the monadological writings, they don't seem to have the role they do, e.g., in the CA; monads, not composite substances, are the building-blocks of the world in those texts.

A second difference involves the notion of matter. In §II I argued that in the period of the CA and the CM, Leibniz seems to recognize the existence of the Scholastics' primary matter, a constituent of corporeal substance distinct from the form, that with the form goes to constitute the corporeal substance. But in the later view, only the forms remain: everything there is is mental in nature, and everything that there is is built up out of these minds, these monads, these "true atoms of nature." Consequently, in the later view, even the "matter" associated with a form must be reinterpreted mentalistically; secondary matter becomes a colony of minds, and, as I earlier noted, primary matter, when recognized at all, becomes an aspect of monadic perception.

Although the two positions are genuinely different, at the same time I don't want to overstate the differences between them, or suggest that Leibniz's thought is in a deep sense inconsistent. What I have called the Aristotelian and the monadological views seem to be two different ways of working out what is, at root, a single position. In both, what is basic to metaphysics is the notion of a substance or unity, and in both, the world of bodies is resolved into a world of organisms, entities composed of souls and bodies. In a way, the two views can be construed as differing only on matters of detail, whether the organisms found everywhere in body are themselves substances, whether it is organisms or souls that are the basic building-blocks of the world, whether we need to appeal to any metaphysical principle over and above incorporeal substance or soul. On these questions the Aristotelianism of the CA seems to represent one extreme, and the idealism of the *Monadology* the other. (From this it should be evident why I think that it is wrong to insist that the metaphysics of the CA *just must* be interpretable in terms of the monadology picture; both can be seen as developments of the same philosophical intuitions without, thereby, being somehow identical or intertranslatable). But given the extent of the agreement between these two pictures, it is not surprising to find Leibniz vascilating on these finer questions over a long period. It is for this reason that it is unrealistic to expect there to be a neat temporal boundary between the one view and the other, as some commentators have claimed to find.[129] Nor should we be disconcerted to find Leibniz making clearly idealistic remarks sandwiched between periods in which the Aristotelian picture seems to predominate, as I shall later point out.[130] The Aristotelian picture is not, I think, a well-defined state through which Leibniz passed on his way to the later idealism of his *Monadology* but a recurring strand in this thought, a strand that Leibniz himself probably didn't distinguish as clearly as I have been trying to do. This conception of Leibniz's development suggests an important historical project, carefully tracing out this Aristotelian strain in all of its manifestations, from its earliest appearances to its latest, as it weaves its way through the Leibnizian corpus. But for the purposes of my project in this paper, in which I shall attempt to connect Leibniz's Aristotelian metaphysics with his conception of the foundations of physics, a more modest investigation will suffice. It is not always easy to tell when Leibniz is talking as an Aristotelian and when he is talking as an idealist; the two pictures are close enough -- variations, as it were, on a single theme -- that much of what Leibniz says could easily be fit into either mold. But there is good reason to believe, I think, that the Aristotelian picture so prominent in the mid and late 1680's persists as a live option throughout the decade of the 1690's and into the next.

The plausibility of this claim is supported by the numerous passages from the correspondence of the 1690's where Leibniz mentions his exchanges with Arnauld, apparently endorsing the positions he took in the CA.[131] At one point, in a letter to Simon Foucher from 1695, Leibniz even considers publishing the entire CA along with the then about to be published *New System of Nature...*, suggesting that the position taken there and the position taken in the CA are the same.[132] But the best evidence for the claim that the Aristotelian picture persists is the rich assortment of passages from writings in the '90s and into the next decade where Leibniz expresses positions on matter, form, and substance similar to those expressed in the CA. In §II we saw in the March 1690 letter to Fardella a recognition that there must be primary matter in bodies, something over and above the form or soul. And in more general terms, the metaphysical picture that Leibniz relates to Fardella is very close to the account given in the CA. As in the CA, one of Leibniz's central concerns is to ground the reality of body in the existence of genuine substances. Leibniz writes:

> A body is not a substance but an aggregate of substances, since it is always further divisible, and any given part has another part *ad infinitum*....Therefore, besides body or bodies, there must be substances to which true unity belongs. For, indeed, if there are many substances, then it is necessary that there must be one true substance....Hence, unless there were certain indivisible substances, bodies would not be real, but only appearances or phenomena, like the rainbow, having eliminated, indeed, every foundation on which they can be composed.[133]

Leibniz is clear here that the substances that ground the reality of bodies are the corporeal substances, the rudimentary organisms, so much in evidence in the CA:

> So in a fish pond there are many fishes, and the liquid in

> each fish is again, like a certain kind of fish pond which, as it were, contains other fishes or animals of their own kinds; and so on to infinity. And therefore, there are substances everywhere in matter, as points are in a line.[134]

It is possible that the "substances everywhere in matter" are just souls. But given the context, it is more likely that the substances he has in mind are organisms, fishes and other smaller animals.

The Aristotelian picture of the CA is also clearly in evidence in writings surrounding the *New System of Nature* of 1695. The *New System* is a difficult work to interpret on this question, and contains some passages that are very suggestive of the later monadology picture.[135] But in an earlier draft of the *New System* Leibniz seems quite clearly to be working within the same picture we found in the CA. Leibniz writes:

> It is necessary that everything be full of those kinds of things, which contain in them a true principle of unity which is analogous with the mind [*ame*] and which is joined in a manner with an organized body, otherwise we would find no substances in matter, and bodies would be only phenomenal and like well regulated dreams.[136]

This same picture is also found in some pieces written just after the *New System*, in Leibniz's responses to Simon Foucher's published critique of that work. In his (unpublished) remarks on Foucher's piece, Leibniz notes that "in actual substantial things, the whole is a result or assemblage of simple substances, or better, of a multitude of real unities."[137] When, later in the same paragraph, he expands on this theme, it is clear that the unities he has in mind are not like minds, but like animals. Leibniz writes:

> In realities where only divisions actually made enter, the whole is only a result or an assemblage, like a flock of sheep. It is true that the number of simple substances that enter into a mass [*masse*] however small it might be, is infinite, since leaving aside the soul which produces real unity in the animal, the body of the sheep (for example) is actually subdivided, that is to say, it, again, is an assemblage of invisible plants or animals, composite in the same way, taken aside from that which also produces their real unity. And although this proceeds to infinity, it is obvious that in the end, all reduces [*revient*] to these unities, and the rest or the results are only

> well-founded phenomena.[138]

Primary matter does not seem to come up explicitly in these texts. But in other respects, they seem similar enough to the letters he wrote to Arnauld some seven or eight years earlier to suggest that the position is basically the same.

Leibniz's Aristotelianism is also evident in a number of letters dating from 1698 and 1699. For example, in a letter to John Bernoulli from September of 1698, Leibniz emphasizes that bodies are made up of corporeal substances analogous to animals:

> You ask me to divide a portion of mass into the substances of which it is composed for you. I respond, there are as many individual substances in it as there are animals or living things or things analogous to them in it. And so I divide it in the same way one divides a flock or a fish pond....You ask how far one must proceed in order to have something that is a substance and not [a collection of] substances. I respond, until a thing without subdivision is displayed, and each such thing is an animal.[139]

And in letters to Thomas Burnett from that same period, Leibniz emphasizes the difference between matter and form. Leibniz writes in early 1698:

> My opinion is thus that matter is only something essentially passive. Thought and even action cannot be modes of it, but only of the complete corporeal substance, which is completely constituted by two constituents, the active principle, and the passive principle, of which the former is called form, soul, entelechy, primitive force, and the second called primary matter, solidity, resistance.[140]

The distinction between matter and form comes out even more clearly in a passage Leibniz wrote to Burnett in 1699 (?), which appeals to the distinction between primary and secondary matter discussed earlier. Leibniz writes:

> In bodies, I distinguish corporeal substance from matter, and I distinguish primary matter from secondary. Secondary matter is an aggregate composed of many animals. But each animal and each plant is also a corporeal substance, having in itself the principle of unity, which makes it truly a substance, and not an

> aggregate. And this principle of unity is that which one calls a soul or indeed something analogous to a soul. But leaving aside the unity, the corporeal substance has its mass [*masse*] or its secondary matter, which is again an aggregate of smaller corporeal substances, and this goes to infinity. However primitive matter [*matière primitive*] or matter considered in itself, is that which one conceives to be in bodies when all principles of unity are put aside, that is, that which is passive, from which arises two qualities: resistance, and inertia [*resistentia et restitantia vel inertia*].[141]

In the letters to Burnett, primary matter is discussed in its connection with passivity. In other writings of these years, primary matter is also linked with extension and mass, and in this respect is in contrast with the non-extended form. Thus, in "On Nature Itself," published in September 1698, Leibniz says that "it is in this passive force of resistance, which involves impenetrability but something more, that I locate the concept of primary matter or mass [*molis*] which is everywhere the same in body and is proportional to its magnitude."[142] Similarly, Leibniz writes to DeVolder in March/April of 1699:

> Thus the resistance of matter contains two factors: impenetrability or antitypy, and resistance or inertia. And since they are everywhere equal in body, or proportional to its extension, it is in them that I locate the nature of the passive principle or of matter, even as I recognize in active force...the primitive entelechy or in a word something analogous to the soul.[143]

Leibniz's commitment to this Aristotelian picture in these years is underscored by the definition of the recently introduced technical term "monad" he gives John Bernoulli in the September 1698 letter from which I quoted earlier. Leibniz writes:

> What I call a complete monad or individual substance [*substantia singularis*] is not so much the soul [*anima*] as it is the animal itself, or something analogous, a soul or form endowed with an organic body.[144]

This conception of the monad, the basic unit of Leibniz's metaphysics, is what he seems to have in mind in the following passage from "On Nature Itself," the work in which the technical term "monad" is first introduced publicly, published in the same month the previous passage was written to Bernoulli:

> Since these activities and entelechies cannot be modes of primary matter, or mass [*molis*], something essentially passive,...it can be concluded that there must be found in corporeal substance a primary entelechy or first recipient of activity....It is this substantial principle itself which is called *soul* in living beings and *substantial form* in other beings, and insofar as it together with matter constitutes a true unity, or a unity in itself [*unum per se*], it makes up [*facit*] what I call monad. For if these true and real unities were dispensed with, only beings through aggregation would remain; indeed it would follow that there would be left no true beings within bodies.[145]

And finally, the position appears prominently in writings from 1703-1704. In a letter from March 1703, Isaac Jaquelet complained to Leibniz:

> [Matter], you say, is a substance which acts and which is passive [*qui agit et qui souffre*]. But it appears to me, Sir, that to speak accurately and philosophically, to act and to be passive are two contradictory things which cannot be found in a single subject. Consequently, this definition appears rather to entail a division of substance into two sorts, into substance which acts, that is the mind [*l'Esprit*], and substance which is passive, that is the body.[146]

Leibniz answers:

> First of all, matter (I understand here secondary matter or mass) is not a substance, but substances, like a flock of sheep or a pond full of fish. I count as corporeal substances only natural machines which have souls or something analogous, otherwise there would be no true unity. Every created substance acts and is passive [*agit et patit*] (there is nothing contradictory in this), and I am of the opinion that none of them is separated from matter....However, we are perhaps in fundamental agreement, Sir, that the mind [*l'esprit*] acts and that matter is passive, since in every corporeal substance I conceive two primitive powers [*puissances primitives*], that is the entelechy or primitive active power, which is the soul in animals, and the mind in man, and which is, in general terms, the substantial form of the ancients, and then the primary matter or primitive

> passive power which provides resistance. Thus it is properly the entelechy which acts, and the matter which is passive, but the one without the other is not a complete substance.[147]

Leibniz's answer suggests both the aggregate argument as found in the CA, as well as a genuine distinction between form or soul and primary matter.

The metaphysics of the CA is also clearly in evidence in the *New Essays*, on which Leibniz worked in 1703-1704. There, for example, one finds Leibniz claiming, as he did nearly twenty years earlier, that the reality of bodies is grounded in living things, things with souls and bodies:

> [Matter] should not be regarded as 'one thing', or (in my way of putting it) as a true and perfect monad or unity, because it is only a mass [*amas*] containing an infinite number of beings....[I] attribute perception to all of this infinity of beings: each of them is like an animal, endowed with a soul (or some comparable active principle which makes it a true unity), along with whatever the being needs to be passive and endowed with an organic body. Now, these beings have received their nature which is active as well as passive [i.e. have received both their immaterial and their material features] from a universal and supreme cause....[148]

These animate beings seem, for Leibniz of the *New Essays*, to be genuine substances, organic bodies unified by souls or the like:

> Nevertheless, it is good to recognize the difference between perfect substances and the assemblages (or aggregates) of substances which are substantial entities put together either by nature or by human artifice....[P]erfect unity should be reserved for animate bodies or bodies endowed with primitive entelechies; for such entelechies bear some analogy to souls and are as indivisible and imperishable as souls are.[149]

And finally, in the *New Essays* Leibniz recognizes primary matter as something in some sense distinct from form. Leibniz writes:

> Thus it is not as useless as you think to reason in general physics about primary matter and to determine its nature -- whether it is always uniform, whether it has any properties other than impenetrability (in fact I have shown, following Kepler, that it also has what could be called *inertia*) and so on

> -- despite the fact that it never occurs naked and unadorned; just as it would be permissible to theorize about pure silver even if we had never found any and had no methods for purifying silver.[150]

Few of these passages I have cited are entirely unambiguous; it is possible that almost every one could be given an idealistic reading, just as one could, perhaps, jam the CA itself into that procrustean bed. However, it seems plausible that the view of the world that Leibniz articulated in response to Arnauld's acute questioning, the emphasis on the corporeal substance as the basic building block of the world, the recognition of something over and above souls or forms, persists through to at least 1704 or so. But I should emphasize once again that the Aristotelianism I have been mainly concerned with is not the *only* strand of thought between 1686 and 1704. The more idealistic strand in Leibniz's thought is suggested, for example, in certain writings from 1695 and 1696; it surfaces again in 1700, 1702, and 1703, before finally *appearing* to win out in roughly 1704 or 1705.[151] Even the *New Essays* which, I have argued, reflects the Aristotelianism of the CA, shows traces of a more idealistic metaphysics.[152] But as unclear as its boundaries are, and as mixed up as it is with other strains of thought, it is fair to say that Leibniz's Aristotelianism as I have developed it in the previous sections remained a live option for Leibniz for at least fifteen years after its clear articulation in the CA.

This concludes what I have to say about Leibniz's Aristotelian metaphysics. In succeeding sections I would like to focus in on this strain of Leibniz's thought and argue that it is in terms of this conception of the world that much of Leibniz's most serious work on the foundations of physics is best understood. But before entering into this project, I would like to pause and reflect a bit on the study of Leibniz's philosophy.

Recent commentators on Leibniz have tended to make two substantive assumptions about his philosophy, that it is linear in the sense of deriving in an almost deductive way from some one set of basic commitments, and that, following the discoveries of the early 1680's or so, as recorded in writings like the DM, CA, and the important "First Truths," Leibniz's doctrine remains essentially unchanged through the rest of his long philosophical career. Both assumptions are explicitly made in Bertrand Russell's monumental and enormously influential *Philosophy of Leibniz*, a work that more than eighty years after its publication still shapes many commentators' view of Leibniz's thought. For Russell, "what is first of all required in a commentator is to attempt a reconstruction of the system which Leibniz should have written -- to discover what is the beginning, and what the end, of his chains of reasoning, to exhibit the interconnections of his various opinions, and to fill in from his other writings the bare outlines of such works as the *Monadology* or the *Discours de Métaphysique*;"[153] the project is, for Russell, to write the systematic work, the *Philosophia Leibniziana demonstrata more geometrico* that Leibniz himself never got around to writing.[154] For Russell, of course, and for most Leibniz scholars writing after Russell and his contemporary Louis Couturat, author of the equally important *Logique de Leibniz*, the beginning of Leibniz's mature philosophy, from its purported first appearance in the 1680's on, is his logic: "No candid reader of [Couturat's] 'Opuscules,'" Russell writes in the preface to the 1937 edition of his book, "can doubt that Leibniz's metaphysic was derived by him from subject-predicate logic. This appears, for example, from the paper 'Primae Veritates' [1680-86?]..., where all the main doctrines of the 'Monadology' are deduced with terse logical rigour...."[155] Much Leibniz scholarship in the years that followed Russell's and Couturat's work has been shaped by these assumptions, the picture of a unified Leibnizian system, logic writ large, maintained virtually without change for over thirty years.

I believe that neither of these assumptions is tenable. While I cannot argue the case here, I think that is wrong to see Leibniz's thought as deriving from his logic, either as a historical or a philosophical claim. This is not because logic wasn't important to Leibniz; it was, and was a source of many arguments and philosophical positions. Rather, I would claim, Leibniz's philosophy doesn't derive from his logic because it doesn't *derive*, strictly speaking, from any one source at all. Leibniz's philosophy is not, I think, a linear argument, with a beginning, middle, and end, but a complex of interrelated and mutually reflecting positions, principles, and arguments; one ought not to ask which came first, the logic, metaphysics, physics, the theology, or mathematics, or moral theory, but rather one ought to see how these domains are interconnected in Leibniz's thought, which is, to use Michel Serres' apt image, more like a net than a chain.[156]

But the business of these last few sections bears somewhat more directly on the second interpretative assumption, the assumption of an unchanging Leibnizian philosophy. If my reading of the DM, the CA, and some of the other texts I cite is correct, then it must be the case that Leibniz's thought underwent some significant changes between its formulations in the 1680's, and the apparently canonical formulation of his last thought in the *Monadology*. But we must be careful here not to substitute a new myth for the old one. Leibniz himself recognized the importance of chronological considerations in the study of a philosopher's work. He wrote:

> [I]t is good to study the discoveries of others in a way that allows us to see the source of their discoveries....And I would like it for authors to give us the history of their discoveries and the process by which they arrived at them. When they do not do so it is necessary that we try to guess about it, the better to profit from their works.[157]

Leibniz himself was quite obliging on this score; in numerous places he gives the readers or correspondents thumb-nail intellectual autobiographies, sketches of the important steps that he went through, from his earliest studies to the position that he is at that time advancing.[158] The picture one gets in these writings is that of an orderly linear development, from Leibniz's earliest acquaintance with the philosophy of the schools, through his infatuation with atomism and mechanism, and on to his later positions. But however that may help us to understand and appreciate Leibniz's later thought, that is not the way it actually happened. A more careful study of the texts cited in this section (and later texts as well) would show, I think, something much more interesting and rich, a complex interplay of positions, Aristotelianism and idealism in many versions with many variants weaving their way toward the monadological metaphysics that Leibniz appears to settle on in his last years.[159] This is not to say that the two myths, Russell's myth of a static Leibnizian system of philosophy and Leibniz's own myth of a continuous progress are entirely mistaken. I think that below the surface flux of hesitations, changes of mind, there may be some very deep commitments, commitments about organism and mentality, unity and substantiality, force and activity, and the basic intelligibility of the world, commitments he came upon and worked out in a gradual way, commitments that shape and limit what appear to us as swings in his mature thought. The tension I have noted between Aristotelianism and idealism in Leibniz's thought in the '80s and '90s may turn out to be different ways of working out these basic commitments, as I suggested earlier. This suggests that the project is one of separating out what persists, more or less, and differentiating it from the different ways the basic commitments are realized at different times; it is this rich and living philosophy, and not Rusell's static reconstruction of the system Leibniz should have written or Leibniz's own sanitized versions of his development that the commentator should seek to capture. But this is a project that goes far beyond the scope of this paper.[160]

V. Corporeal Substance and the Biological Foundations of Dynamics

I have argued that in the 1680s and the 1690s an important strand of Leibniz's thought was the picture of a world of corporeal substances, organisms understood on an Aristotelian model, unities of form and matter, a world different from the world of the *Monadology*. But this period is *also* the period in which Leibniz was most active in developing his mature physics. The DM was written the same year Leibniz's important "Brief Demonstration" appeared in print, touching off the lively debates about *vis viva* and quantity of motion that were to continue for some years; Leibniz penned the massive *Dynamica de Potentia et Legibus Naturae corporeae*, the most complete account of his dynamics and the source for the *Specimen Dynamicum* of 1695 in the shadow of his CA and only shortly before the illuminating letter to Fardella I cited earlier. In fact, *after* 1700 or so Leibniz seems to show hardly any interest at all in technical work in physics. It would be strange indeed if the account of body and the make-up of the material world that so attracted Leibniz at this time wasn't *somehow* reflected in his writings on physics. And I think it was. In fact, I claim, it is in terms of this Aristotelian picture that we can best understand Leibniz's conception of the metaphysical foundations of physics at this time: the living things, organisms, corporeal substances which so fascinated Leibniz in this period are, I claim, *the very hooks on which Leibniz hangs his new science of dynamics.*

In order to demonstrate the connection between Leibniz's dynamics and the Aristotelian strain in his metaphysics, we must begin by saying something about Leibniz's approach to physics. There is not space in this essay to develop Leibniz's complex program for physics, for which he coined the term "dynamics." But something must briefly be said about Leibniz's conception of force and the role it plays in his physics, for it is force that defines Leibniz's dynamical approach to physics.

Leibniz's conception of force is, perhaps, best understood as a reaction against certain aspects of the Cartesian conception of physics, a conception that was to dominate much of 17th C physics and remain influential into the next century. Descartes' general program in physics was what has come to be called mechanism, the attempt to explain all features of the material world in terms of the shapes, sizes, and motions of the bits of homogeneous matter, generic material substance, that, it is claimed, makes up the material world.[161] Basic to this project, of course, are the laws of motion and impact, the laws that specify the microscopic behavior of the corpuscles that is supposed to account for the macroscopic behavior of sensible bodies. Descartes' strategy for deriving the laws of motion and impact is highly ingenious. Descartes begins his most careful account of the laws of motion, the one given in part II of his *Principia Philosophiae*, with a careful distinction between motion and its cause. Motion, strictly speaking, is just the translation of a body from one region to another, and, Descartes claims, is just a mode of extension on a par with shape or size.[162] But the notion of motion is not by itself sufficient for deriving the laws that bodies in motion obey, for Descartes. For this we must turn away from motion, the effect, and examine its cause, God.[163] Descartes' God is continually recreating the world from moment to moment, keeping it in existence by an activity that is no different in nature from the activity by which he originally created it. God emerges on this picture as the principal cause of motion since all motion must result from His moment by moment reshuffling of the positions bodies have with respect to one another. And just as God is the ground of motion for Descartes, He is also the ground of the *laws* of motion; it is because of God's attributes, the *kind* of being God is, that the laws of motion are what they are. So, Descartes claims, it is because God is immutable and constant in his operation that he maintains the same quantity of motion in the world, that a body in rectilinear motion will persist in that motion, that bodies colliding with one another will behave in the way Descartes thinks that they will.[164]

Leibniz's physics has much in common with Descartes'. As we shall see below, Leibniz, like Descartes, was a mechanist, and believed that everything in the physical world could be explained in terms of size, shape, and the laws of motion. Leibniz, like Descartes, also drew a clear distinction between motion (and, more generally, the characteristic behavior a body has, its impenetrability, inertia, etc.) and the *cause* of motion or other behavior. And finally, Leibniz, like Descartes, gave God a crucial role in the derivation of the laws of motion. But there is one central feature of Descartes' account that Leibniz firmly rejects. It is crucial to the Cartesian derivation of the laws of motion that God is the agent who, in his continual recreation of the world, shuffles bodies about; it is because of this and God's nature that bodies in motion obey the laws they do. But for Leibniz, while God is the *ultimate* cause of motion, just as he is the ultimate cause of *everything* in the world, motion and any other features of the behavior of bodies must derive from something *in the bodies themselves.* Leibniz offers a number of considerations in favor of this position. One seems connected to the argument of DM8 and the claim that substances must contain that by virtue of which their predicates apply. Thus Leibniz writes in the *Specimen Dynamicum* of 1695:

> ...I freely admit that all things arise by a continual creation from God; yet I think that there is not natural truth in things whose reason derives immediately from divine action or will, but that God has always put into things themselves some properties by which all their predicates can be derived.[165]

It might be complained here that this holds only for genuine substances, and that if bodies are not substances (or do not contain substances) then there is no reason why they should contain the grounds of their predicates. But, Leibniz argued, if bodies were not somehow made up of substances, that would lead to certain evident absurdities; if bodies did not contain substances, and did not contain the grounds of their predicates or, what Leibniz took to be equivalent,[166] the sources of their actions, then the world of bodies would be a mere mode of God. Leibniz writes in *On Nature Itself*:

> ...[T]he very substance of things consists in the force of acting and being acted upon; hence it follows that no enduring thing can be produced if no force that endures for some time can be impressed upon it by the divine power. Then it would follow that no created substance, no soul would remain numerically identical, and hence that nothing would be conserved by God, but everything would reduce to certain evanescent and flowing modifications or phantasms, so to speak, of the permanent divine substance. And, what reduces to the same thing, God

> would be the nature or substance of all things -- a doctrine of most evil repute, which a writer who was subtle indeed but irreligious, in recent years imposed upon the world....[167]

This *reductio ad Spinozam* was sufficient criticism from Leibniz's point of view. But there is one last argument, perhaps Leibniz's best: a world of substances active in themselves is simply a world more fitting for a perfect and omnipotent God to have created than the Cartesian world of inert bodies moved by God. Leibniz writes, again in "On Nature Itself":

> For I ask whether this volition or command, or, if you prefer, this divine law once established has bestowed upon things only an *extrinsic denomination* or whether it has truly conferred upon them some created impression which endures within them..., an *internal law* from which their actions and passions follow....For since this command in the past no longer exists at present, it can accomplish nothing unless it has left some subsistent effect behind which has lasted and operated until now....It is not enough, therefore, to say that in creating things in the beginning, God willed that they should observe a certain law in their progression, if his will is imagined to have been so ineffective that things were not affected by it and no durable result was produced in them. And, indeed, it conflicts with the concept of the divine power and will, which are pure and absolute, for God to will and yet to produce or change nothing by his willing, for Him always to act but never to achieve and not to leave behind any work or any accomplishment.[168]

The cause of motion and its law must, then, be not only in God, but in the bodies themselves.[169]

That something in bodies that accounts for their behavior cannot be merely extension or motion, Leibniz claims. Leibniz has a number of arguments to this conclusion, but one, an argument he rehearses and refers to in numerous places, is especially lucid. In a short version, given in DM21, it reads as follows:

> For if there were nothing in bodies but extended mass, and nothing in motion but change of place, and if everything should and could be deduced solely from the definitions of these by geometric necessity, it would follow...that the smallest body, in colliding with the greatest body at rest, would impart to it its own velocity, without losing any of this velocity itself; and it

> would be necessary to accept a number of other such rules which are entirely contrary to the formation of a system.[170]

What Leibniz has in mind is something like this. Let us suppose that bodies are only extended, and, to allow for the very possibility of collision, impenetrable. Now, in such bodies, there is nothing from which new motion can arise and nothing that can actively oppose the acquisition of new motion. Consequently, if a body B, at rest, is hit by a body A, in motion, then A's motion will continue unimpeded, and, since B is impenetrable, A will carry B along with it. In such a circumstance, A and B will move off together after the collision at A's pre-collision velocity; *and this holds no matter how small A and how large B are.* Thus, if A and B are just extended and impenetrable, then "the smallest body, in colliding with the greatest body at rest, would impart to it its own velocity, without losing any of this velocity itself."[171] More generally, if all there were in bodies was extension and impenetrability, Leibniz argues, the correct laws of impact would simply be the law of the composition of velocities.[172] But that would be absurd; it is not what we observe to happen in the world, nor is it consistent with certain metaphysical principles.[173] So, Leibniz concludes, there must be something more to bodies than extension and impenetrability; extension and impenetrability alone cannot account for why bodies behave the way they do.[174]

So, Leibniz thinks, if we are to avoid Cartesianism and place the cause of motion and its laws in bodies themselves, then we must reject the Cartesian ontology and see in bodies something over and above extension and motion. This something is what Leibniz prefers to call *force* or sometimes power in his dynamical writings, though it has important connections with other notions, as we shall later see. Leibniz thus writes in the *Specimen Dynamicum* in concluding a lengthy presentation of the argument we have just discussed:

> I concluded, therefore, that besides purely mathematical principles...there must be admitted certain metaphysical principles...and that a certain higher and so to speak, formal principle must be added to that of material mass....Whether we call this principle form, entelechy, or force does not matter provided that we remember that it can be explained only through the concept of forces.[175]

Leibniz's concept of force involves two important pairs of distinctions, the distinction between active and passive forces, and between primitive and derivative forces. So, in all, Leibniz recognizes four principal varieties of force, primitive active and passive, and derivative active and passive. While the distinction between primitive and derivative forces will be important to us later, for the moment I would like to focus in on the distinction between active and passive forces and their relations to the observable behavior of bodies.

Active force is the force that, for Leibniz, is manifested in motion, both actual motion and instantaneous acceleration. Leibniz writes in the *Specimen Dynamicum*:

> [Active] force is of two kinds: the one elementary, which I also call *dead* force, because motion does not yet exist in it but only a solicitation to motion such as that of...a stone in a sling even while it is still held by the string; the other is ordinary force, that which is connected with actual motion, which I call *living* force. An example of dead force is centrifugal force, and likewise the force of gravity or centripetal force, also the force with which a stretched elastic body begins to restore itself. But in impact, whether this arises from a heavy body which has been falling for some time, or from a bow which has been restoring itself for some time, or from some similar cause, the force is living and arises from an infinite number of continuous impressions of dead force.[176]

This suggests that active force is to be connected with velocity and acceleration, living active force with velocity, and dead active force with acceleration.[177] Passive force is something quite different. Leibniz writes, again in the *Specimen Dynamicum*:

> [Passive force] brings it about that one body is not penetrated by another, but opposes an obstacle to it, and is at the same time possessed of a kind of laziness [*ignavia*] so to speak, or a repugnance to motion, and so does not allow itself to be set in motion without somewhat breaking the force of the body acting upon it.[178]

So passive force is connected with two features of the behavior of bodies: their impenetrability, and what we have come to call their inertia, following Newton.[179]

It is important to note that while active force is associated with motion (velocity and acceleration), and passive force is associated with resistance (impenetrability and inertia), Leibniz is very careful to distinguish these forces from the observable behavior of bodies with which they are associated. The distinction between living active force and the motion that accompanies it is one of the important lessons of the important "Brief Demonstration," published in 1686, the year in which the DM was written. In that argument, rehearsed in DM 17 and in many other writings of the 1680s and 1690s, Leibniz argues that if the force in question is proportional to the ability a body has to accomplish an effect (lifting itself or another body a certain vertical distance), then while force is *connected* with motion, they are distinct; force must be proportional not to velocity, but to the square of velocity, so that if we double the velocity of a body we quadruple its force.[180] More generally, Leibniz identifies active force, both living and dead, as the *cause* of motion observed in bodies. It is, in fact, by appeal to active force, the cause of motion, that we can break the relativity of motion considered merely as change of place, and assign motion and rest to bodies in a non-arbitrary way. Thus Leibniz writes in DM 18, in a passage that immediately follows his exposition of the "Brief Demonstration..." argument:

> For considering only what it means narrowly and formally, that is, a change of place, motion is not something entirely real; when a number of bodies change their position with respect to each other, it is impossible, merely from a consideration of these changes, to determine to which bodies motion should be ascribed and which should be regarded at rest, as I could show geometrically....But the force or immediate cause of these changes is something more real, and there is a sufficient basis for ascribing it to one body rather than to another. This, therefore, is the way to learn to which body the motion preferably belongs. Now, this force is something different from size, figure, and motion, and from this we can conclude that not everything which is conceived in a body consists solely in extension and its modifications, as our moderns have persuaded themselves.[181]

While Leibniz is not as explicit in the case of passive forces, there is reason to believe that here, too, he was interested in distinguishing passive force from the behavior in bodies that it causes. In the *Specimen Dynamicum*, as we have already seen, passive force is carefully characterized as that which "*brings it about* [*fit*]" that bodies are impenetrable and exhibit inertia. That is, passive force is that by virtue of which, in a given impact, say, one body is prevented from penetrating another and that by virtue of which the one body breaks the force (and thus the motion) of the others; it seems to be the cause of the observed behavior in something of the same way that active force is the cause of motion. Passive force seems also to be, in a somewhat extended sense, the cause of a body's extension, insofar as extension in body results from its passive force. Thus, in a passage from the CA that we have already seen in connection with primary matter, one that echos some further questions to which we shall later turn, Leibniz talks of the "primitive passive power [i.e., force]" of a substance as its matter, claiming that "in this sense matter would not be extended or divisible, although it would be the principle of divisibility or that which amounts to it in the substance."[182] The suggestion seems to be that what is called extension by the physicist is, then, just the diffusion of resistance; it is, properly speaking, a consequence of the property bodies have by virtue of which they resist penetration by other bodies.[183]

So far we have been mainly concerned with the notion of force and, in very general terms, the roles it plays in Leibniz's conception of the physical world. At this point I would like to introduce some of our earlier concerns and begin to draw a connection between the dynamical conception of the world and the Aristotelian metaphysics discussed in the earlier sections of this paper.

Leibniz connects the notion of force in his dynamics with that of substance in his metaphysics in the most general terms. In his "Correction of Metaphysics..." of 1694 he writes, for example, that "the concept of *forces* or *powers*, which the Germans call *Kraft* and the French *la force*, and for whose explanation I have set up a distinct science of *dynamics*, brings the strongest light to bear upon our understanding of the true concept of *substance*."[184] This much is widely known. But it is not so widely known that the substances that Leibniz seems to have in mind here, the substances that he first links to forces are not the later monads, but the corporeal substances of the more Aristotelian world of the CA. For example, in writing to Arnauld in 1687, in a paragraph explicitly concerned with substantial forms and the "corporeal substances endowed with true unity" that they help make up, Leibniz writes of "force which is the cause of the motion, *and which exists in corporeal substance*."[185] Similarly, in the first draft of the "New System..." of 1695, probably written the same year as the "Correction of Metaphysics," Leibniz relates the notion of force that grounds his physics with the notion of form that grounds the Aristotelian picture. Leibniz writes:

> It appears even that in all organic things there is need for something which corresponds to the soul, and which the philosophers have called substantial form, which Aristotle calls first entelechy, and which I call, perhaps more intelligibly, primitive force.[186]

Forces, then, pertain to the *unum per se*, the organisms that in this strand of Leibniz's thought constitute corporeal substances.

This establishes a very general connection between force and corporeal substance. But the connection runs much deeper than that. The last quotation suggests that in order to understand how force is connected to substance, we must understand what primitive forces are, and, presumably, how they differ from derivative forces.

Derivative forces, when introduced in the *Specimen Dynamicum* are identified as the forces that the physicist is most directly concerned with; derivative force is, as Leibniz understands it,

> ...the force by which bodies actually act and are acted upon by each other,...that force which is connected with motion, local motion, that is, and which, in turn, tends to produce further local motion.[187]

Derivative force is, furthermore, the notion of force in terms of which we can frame *laws* of physics:

> To these derivitive forces apply the laws of action, which are not only known to reason but also verified by sense itself through phenomena.[188]

Leibniz relates these derivative forces to primitive forces as follows in the *Specimen Dynamicum*:

> The derivative force...is exercised in various ways through a limitation of primitive [force] resulting from the conflict of bodies with one another.[189]

Derivative force is, thus, a *limitation* of primitive force. This suggests, in the philosophical vocabulary of the 17th Century, that derivative force is a *mode* of primitive force, an *accident* which inheres in primitive force. And this is, indeed, what Leibniz says in other passages. Leibniz writes, for example, in the first draft of the "New System...":

> [I call form or entelechy] the primitive force in order to distinguish it from the secondary [i.e., derivative force], what

> one calls moving force, ***which is a limitation or accidental variation of the primitive force.***[190]

Similarly, Leibniz writes in a letter to Bernoulli from 1698:

> For the rest, if we conceive of soul or form as the primary activity from whose modifications the secondary [i.e., derivative] forces arise as figures arise from the modifications of extension, I believe we shall have satisifed the demands of understanding. There can be, that is, no active modifications of something whose essence is purely passive, because modifications limit rather than increase or add.[191]

And finally, consider the following passage from the Appendix of 1702:

> Active force is twofold, primitive and derivative, that is, either *substantial* or *accidental.*[192]

These passages suggest that derivative forces are to be understood as modes, accidents or the like, modifications of the primitive forces. Given our post-Newtonian conception of force the claim is a bit puzzling. But Leibniz's picture is quite straightforward. *Our* notion of force is that of a physical magnitude, something that can take on different values in different bodies and at different times.[193] Leibniz's primitive forces are not forces like that at all; Leibniz's primitive forces seem to be genuine things, the seat of modifications. A thing doesn't have Leibnizian *primitive* force in the way in which a body in acceleration has Newtonian force, but rather, in the way in which one might talk about the Church as a *force* in society, or Mohammed Ali as a *force* to be reckoned with in the boxing ring. Primitive forces are things, but things capable of exerting particular forces in particular situations. Such particular forces, exerted in particular situations, may be considered modes, accidents, or, in the 17th Century idiom, limitations of the primitive force, just as particular volitions are modes of mind, or particular figures modes of extension. And these modes of primitive force are what Leibniz calls derivative forces, and they are the forces of physics.[194]

And with the distinction between primitive and derivitive force in place, we can see precisely how the notion of force treated in Leibniz's dynamical writings links to the Aristotelian conception of substance outlined in his more metaphysical works: primitive active force *just is* form, soul, entelechy, and primitive passive force *just is* primary matter. Leibniz thus writes in the *Specimen Dynamicum*, in a passage echoed in many other writings:

> Primitive [active] force, which is nothing but the first entelechy, corresponds to the soul or substantial form....Primitive [passive] force of suffering or of resisting constitutes the very thing which the Scholastics call *materia prima*, if rightly interpreted. It brings it about, namely, that one body is not penetrated by another, but opposes an obstacle to it...and so does not allow itself to be set in motion without somewhat breaking the force of the body acting upon it.[195]

It is not unreasonable to suppose that the form and matter at issue here, the form and matter Leibniz identifies with the two varieties of primitive force, are just the form and matter that play such a central role in the metaphysical writings of the '80s and '90s, the form and matter that compose a corporeal substance. In fact, in the important Appendix of 1702 mentioned earlier, Leibniz explicitly links the physical notions of primitive active and passive force to the metaphysical problems of substance and unity that loomed so large some fifteen years earlier in the CA. Leibniz writes:

> Primitive active force, which is called by Aristotle the first Entelechy, generally the form of substance, is another natural principle which, with matter or [primitive] passive force, completes corporeal substance which is, indeed, an *unum per se*, not a mere aggregate of many substances; for there is much difference, e.g., between an animal and a flock.[196]

If the primitive forces are just the form and matter of corporeal substance, then the derivative forces also must pertain to corporeal substance in an obvious way: derivative forces, the forces that give rise to actual motion, are simply modes or accidents of the form and matter that go to make up a corporeal substance. Though not substantial themselves, they nevertheless pertain directly to the corporeal substance, the basic metaphysical unit of Leibniz's world.[197]

We have seen how derivative forces are in corporeal substances as modes of form and matter. But the connection between dynamics and the metaphysics of corporeal substance goes deeper still. I argued earlier that forces are introduced in bodies to replace Descartes' God, an external source of motion, with an internal source for the behavior of bodies. Thus, Leibniz introduces active forces (grounded in form) to account for motion and passive forces (grounded in primary matter) to account for resistance. But in Cartesian physics the connection between God and the motion of bodies is supposed to account for the laws bodies obey; it is because the Cartesian God is immutable and constant in operation that quantity of motion is conserved, that motion persists, that bodies behave the way they do in collision. Leibniz's God, too, has an important role in fixing the laws of physics; as in Descartes, Leibniz's God is the ultimate ground of the laws of nature. Thus Leibniz writes in a reply to Malebranche published in 1687:

> ...[T]he true physics should in fact be derived from the source of the divine perfections. It is God who is the ultimate reason of things, and the knowledge of God is no less the source of science than his essence and his will are the sources of beings....I agree that the particular effects of nature can and ought to be explained mechanically....But the general principles of physics and mechanics themselves depend upon the action of a sovereign intelligence and cannot be explained without taking it into consideration.[198]

But while Leibniz's God is the *ultimate* source of law, He does not carry His own laws out by His direct activity, as Descartes' God does in continuously recreating the world and, in doing so, shuffling bodies about. For Leibniz, God chooses the laws bodies should obey, and then imprints those laws *in the bodies themselves*. Now, for God's law to be in bodies, there must be something *in* bodies to receive and preserve that decree of God's; as Leibniz wrote in "On Nature Itself,"

> For since this command in the past no longer exists at present, it can accomplish nothing unless it has left some subsistent effect which has lasted and operated until now.[199]

Now, a subsistent effect requires a subsistent entity, something that persists and preserves the impression that God made on it, one would think. But in Leibniz's world (the Aristotelian world) it is corporeal substances that persist. This suggests that it is corporeal substances that should be the ground of God's law in the world. But taking this line of reasoning a step further, in corporeal substances it is the soul, form, or entelechy that persists and serves to identify the corporeal substance as the individual that it is, while the body changes (sometimes, as in death, quite radically) around it. This suggests that it is the soul, form, or entelechy that should be the seat of law in the physical world. While this argument is at best implicit in Leibniz's writings, the conclusion is quite clear. Leibniz notes, for example, in DM18 that

> ...the general principles of corporeal nature and of mechanics themselves...pertain to certain forms or indivisible natures as the causes of what appears rather than to the corporeal or extended mass.[200]

Similarly Leibniz notes in the Appendix of 1702 that "only in entelechies...are placed the principles of mechanism, by which all things are regulated in bodies."[201] And finally, in "On Nature Itself" Leibniz writes:

> If...the law set up by God does in fact leave some vestige of him expressed in things, if things have been so formed by the command that they are made capable of fulfilling the will of him who commanded them, then it must be granted that there is a certain efficacy residing in things, a form or force such as we usually designate by the name of nature, from which the series of phenomena follows according to the prescription of the first command.[202]

The laws that Leibniz thinks God imposed on forms seem to be very general metaphysical principles. In the *Specimen Dynamicum*, for example, Leibniz suggests that the laws in question are the law of continuity, the law of action and reaction, and the law of the equality of cause and effect.[203] These metaphysical principles, applied to the derivative forces, the sort of force to which "apply the laws of action," as Leibniz writes in the *Specimen Dynamicum*,[204] have a central role to play in Leibniz's derivation of the laws that govern motion. Since derivative forces are that in bodies which causes motion and the resistance to motion, the metaphysical principles that God has imposed on derivative forces can be used to determine the laws of motion; these metaphysical principles, Leibniz claims, entail precise quantitative laws for bodies in terms of their size (mass) and velocity.[205]

It should be noted that Leibniz's doctrine on law and its relation to corporeal substance and form is filled with obscurity and positions only partially worked out. For one, it is very unclear just *how* the laws are imprinted. Leibniz, for example, writes in the CA that:

> I therefore believe that there are a few free primary decrees capable of being called laws of the universe, which, when linked to the free decree to create Adam, bring about the

consequence....[206]

This suggests that God places the laws of nature explicitly as such in the form of a corporeal substance, places the substance in certain circumstances (the initial condition), and lets things follow. Elsewhere, though, Leibniz talks of the form, the primitive active force, as "the law of the series," and the individual derivative forces as the individual members of the series.[207] This mathematical analogy suggests a different conception of the laws of nature. On that conception God explicitly imprints in the form the *exact* sequence of events that are to occur in the substance, the precise values that the derivative forces are to take. On this conception, the laws of nature seem to be contained only implicitly, as external constraints on the possible sequences that God might have created. Also, there is a puzzle raised by Leibniz's claim that the laws are imprinted *only* on the form, the primitive active force. Presumably there are laws of nature that involve the passive forces. But if the laws are only in the form, where do the laws governing resistance come from? Why does Leibniz think that it is enough to place the laws in the form alone? In these ways, Leibniz's claim to ground natural law in corporeal substance seems more like a program than like a carefully worked out doctrine. But be that as it may, however one might want to resolve the many obscurities in Leibniz's position, it seems clear that the corporeal substance of Leibniz's Aristotelian metaphysics is the ground not only for the derivative forces, the forces of physics, but also for the laws that govern such forces.

The picture that emerges is quite striking and quite strikingly beautiful. In the earlier sections of this paper I outlined an Aristotelian world of substances, form and matter, soul and body, substance in substance and organism in organism to infinity. In this section I have connected that picture with Leibniz's dynamics, and argued that the forces that constitute the central notions of Leibnizian physics are directly connected to the form and primary matter of his Aristotelian metaphysics. Given the connection between corporeal substance and organism in Leibniz's thought, this suggests a kind of reduction of physics to biology: the forces of physics are, ultimately, to be understood on analogy with the activity and passivity of organisms, active forces on analogy with an organism's intentional behavior, that which derives from its soul or form, and passive forces on analogy with the resistance to activity that derives from its body. Physics, then, deals with forces that derive from the souls and bodies of multitudes of tiny organisms, and with the laws that God has imprinted on their souls. But this mad biology makes for rather strange physics. What sense, on this picture, can Leibniz possibly make of the physics of non-organic, inanimate bodies, bodies that have no claim to substancehood on Leibniz's view, bodies that are *usually* taken to constitute the subject-matter of physics?

One is tempted to see a kind of phenomenalism looming up. As I noted earlier, inorganic bodies are, for Leibniz, phenomenal; in this period, at least, they are aggregates of an infinity of organic corporeal substances, and lack the substantial form that is needed to make them genuine individuals. Their unity comes from us, *we impose* it on them. One is tempted to suppose that the physics of such phenomenal bodies must be something altogether different from the quasi-physics that governs genuine corporeal substances.

But this turns out not to be the case. From Leibniz's point of view, the physics of the two domains, corporeal substances and their aggegates, the non-organic bodies of everyday experience, *is exactly the same*; even though there may be an ontological gap between the domain of genuine entities and their aggregates, they share the same physics.

While Leibniz never addresses this problem directly, he places the tools necessary to deal with this question within easy reach. To put it simply, for Leibniz, an aggregate of bodies, animate *or* inanimate, can be treated as if it were a *single* body, whose mass is the sum of the masses of its parts, and whose velocity is that of the center of mass (gravity) of the aggregate. And while this aggregate may lack a *primitive* active force insofar as it lacks a single form that unites its parts, it *has* derivative forces; its derivative passive force will be proportional to its total mass, and its derivative active force will be measured by the total mass and the square of the velocity of the center of mass.[208] These aggregate derivative forces, derived from the genuine derivative forces of the corporeal substances that make up a body, behave in accordance with the same laws that govern the forces in genuine substances.[209] So, from the point of view of *physics*, the aggregate, the inanimate body *can* be treated *as if* it were a genuine corporeal substance. The physics of corporeal substances thus extends directly to the physics of corporeal aggregates. But, of course, this physics of phenomenal aggregates is, in an obvious way, *parasitic* on the physics of corporeal substances, the *true* science of dynamics. For, as Leibniz emphasizes over and over, if inanimate bodies *weren't* made up of such corporeal substances, then they could have *no forces at all* and would obey laws quite different from the ones they obey.[210]

And at this point we can return to the question I raised at the very beginning of this paper, the question about the status of physics in the Leibnizian system. I talked earlier about a common reading in accordance with which physics comes out as a science of the phenomenal, a science of things as they appear to us rather than a science of things as they really are. In a sense this is true for the texts that we have been considering in this section. Insofar as physics is concerned with motion and the laws that govern motion, it does, at least in part deal with something that Leibniz considers phenomenal in a well defined sense, and something that he considers less real than the forces which cause motion and whose laws determine, in that way, the laws that govern motion.[211] If the laws of mechanism are just the laws that govern motion, then, in whatever sense motion is phenomenal, the laws of mechanism, and, perhaps, the whole mechanical philosophy will be phenomenal as well, an account of the world as seen from our imperfect point of view. And insofar as physics is concerned, at least in part, with inanimate bodies, at least one of its objects is straightforwardly phenomenal, aggregates of genuine substances, whose unity derives from us. But there is a deeper sense in which this common reading is a serious misrepresentation of the physics of Leibniz's middle years. While motion may be phenomenal, and its science in some sense a science of the apparent, forces are at the foundation of Leibniz's conception of what is real in the world, and their science, dynamics, is as real as the forces it treats. And while dynamics may apply to the phenomenal aggregates of substances that constitute inanimate bodies, it is, properly speaking, the real science of the modes of corporeal substance, that which is real in the fullest sense Leibniz recognizes. Leibniz's dynamics, the core of his program for physics, is directly grounded in a world of corporeal substances, and the forces dynamics treats, the derivative active and passive forces, are themselves the momentary states of form and matter, the constituents of corporeal substance. It is difficult to see what more a realist in physics could ask for.

But this picture does not persist throughout Leibniz's career. With the apparent domination of the idealistic picture of the *Monadology* comes an importantly different conception of physics, one much closer to what I have called the common reading than the conception found in the texts we have been examining in this section. Leibniz's physics in his middle years,

I have argued, is grounded in forces that are modes of corporeal substances, the basic constituents of his world. But with Leibniz's later focus on mind-like monads as the ultimate constituents, with his rejection of primary matter as a principle distinct from the mental, and with the problematic status that corporeal substances later have, the core of Leibniz's physics seems to undergo a fundamental transformation. Put most simply, the derivative forces which dynamics treats, once linked directly to substances, now lose their direct link to that which is real. With the replacement of the corporeal substance as the basic building block with the monads of the *Monadology,* derivative force becomes, now, a feature of *aggregates of substances and of aggregates alone.* Derivative forces retain their connection to bodies; but bodies, all bodies, including animate bodies, are resolved into collections of monads. And insofar as such aggregates are considered phenomenal in a well defined sense, even dynamics must relinquish its claim to be a science that pertains to the world as it really is, and must become a science of the phenomenal, a science that pertains to the world as it appears to us, a world shaped at least in part by us. Thus Leibniz writes to DeVolder in 1704-1705:

> I do not eliminate body, but bring it back to what it is, for I show that corporeal mass, which is believed to have something over and above simple substances, is not a substance, but a phenomenon resulting from simple substances which alone have unity and absolute reality. I relegate derivative forces to phenomena, but I believe that it is manifest that primitive forces can only be the internal tendencies of simple substances....[212]

And similarly, Leibniz writes to DesBosses in 1706:

> ...From a multiplicity of monads results secondary matter, along with derivative forces, actions, passions, which are only entities through aggregation, and therefore semi-mental, like the rainbow and other well-founded phenomena.[213]

This phenomenalistic interpretation of derivative force and the status that it entails for the science of dynamics is, to be sure, a genuine Leibnizian position. But, it must be emphasized, this represents a fundamental change from the thought of Leibniz's middle years. In the '80s and '90s it's the corporeal substance that's the thing, the center of Leibniz's thought and the basic building-block of his world. And it is the corporeal substance that provides a real foundation for his science of dynamics.

VI. Scholasticism and Mechanism in Leibniz's Physics

In the previous section I argued that in these middle years, Leibniz's physics is firmly grounded in reality, that the world of Leibniz's physics is not the world of phenomena, but rather a world of corporeal substances that are real in the fullest sense. This settles the principal question that I raised at the beginning of this paper, I think. But the particular way in which Leibniz chooses to ground his physics in a world of Aristotelian substances raises an interesting philosophical question within the historical context of Leibniz's thought. The question, quite simply, is this: having adopted as much of Scholastic metaphysics as he did, how can Leibniz fail to accept Scholastic physics as well?

Let me clarify the question by reviewing a bit of the historical context. Seventeenth Century philosophy was, in large part, a battleground between the old and the new, between the adherents of the Aristotelian philosophy of the schools that had dominated the Latin West since the 13th Century, and the adherents of the new (or newly revived) mathematical and mechanical world view of Galileo, Descartes, and their many minions. For the Scholastics, as I have noted, the world is made up of substances, form united with matter, resulting in a thing that exhibits certain characteristic behavior. Consequently, for the Scholastic scientist (at least as parodied by his mechanist opponents), the behavior and characteristic properties of a body are ultimately explained in terms of its form.[214] Molière's familiar "dormitive virtue" account of opium is, of course, a paradigm example of this mode of explanation. For the mechanist, on the other hand, the manifest properties of bodies are to be explained in terms of the size, shape, and motion of the tiny corpuscles which, it is claimed, make up the bodies of everyday experience. Such an account is explicitly opposed to the explanation of physical phenomena in Scholastic terms, in terms of forms and the innate tendencies that bodies of certain sorts have to rise or fall, to be hot or cold, to reason or to neigh. Consider, for example, the illuminating contrast that Descartes draws in *Le Monde* between the mechanical account of combustion he favors and the Scholastic account he rejects:

> When it [i.e., fire] burns wood or some other such material, we can see with our own eyes that it removes the small parts of this wood, and separates them from one another, thus transforming the more subtle parts into fire, air, and smoke,

> and leaving the grossest parts as cinders. Let others imagine in this wood, if they like, the form of fire, the quality of heat, and the action which burns it as separate things. But for me, afraid of deceiving myself if I assume anything more than is needed, I am content to conceive here only the movement of parts.[215]

For Descartes and his followers, this program in physics was intimately connected with a claim (metaphysical, if you will) about the nature of body. The claim is that mechanism is right and the Scholastic program for science wrong because there are no forms of the sort that the Scholastics envision. Instead, the claim is, bodies are composed of extension and extension alone, and, thus, the only properties that they really have are broadly geometric in nature: size, shape, motion, and, perhaps, impenetrability. And because this is the way bodies are, we must be mechanists all.

Against this historical background, Leibniz's position is not a little puzzling. I have argued at some length that Leibniz accepts a basically Aristotelian account of the make-up of the world, and that such an ontology is what grounds his physics. But -- and here is the puzzle -- the physics that Leibniz thinks he is grounding is *not* the Aristotelian physics of the Schools, but the mechanical philosophy of the Cartesian tradition. Leibniz repeats over and over in unambiguous terms that in physics, he is a modern, that in physics, he is a mechanist, *despite* his endorsement of a Scholastic ontology. "All bodily phenomena can be explained mechanically or by the corpuscular philosophy," Leibniz wrote to Arnauld in a characteristic passage.[216] In fact, Leibniz often argues, the *only* way we can ground mechanism is by *rejecting* the foundations that Descartes so carefully laid, and adopting an Aristotelian foundation for the mechanical philosophy in form and matter; the laws of motion themselves, the basis of any mechanical explanation, require that there be something in bodies over and above extension, something from which force could arise. Thus Leibniz wrote in the DM:

> The general principles...of mechanics [i.e., the laws of motion and force] themselves are...metaphysical rather than geometrical and pertain to certain forms or indivisible natures as the causes of what appears rather than to the corporeal or extended mass. This reflection is capable of reconciling the mechanical philosophy of the moderns with the caution of certain intelligent persons of good will who fear, with some reason, that

> we may withdraw too far from immaterial beings and therefore put piety at a disadvantage.[217]

Such a position must have surprised if not astonished Leibniz's contemporaries. How, a contemporary might ask, with such a metaphysics, can Leibniz avoid rejecting mechanism and reviving the whole godawful mess of Aristotelian science? How does the foundation that Leibniz builds for the mechanical philosophy avoid eliminating it altogether?

Leibniz has a rather ingenious answer to this question. Since the basic unit of the created world is the living organism, soul or form united to matter, one can, indeed, explain its behavior (and, thus, the behavior of the inanimate bodies that it makes up) *modo scholastico*, in terms of form. This may sound strange to the 20th Century Leibniz scholar. But the point I am making is just a restatement of a familiar doctrine of Leibniz's. It is well known that for Leibniz, every substance has a complete individual concept (CIC) which contains all of the properties, past, present, and future that pertain to that substance. And it is well known that this CIC is grounded in certain non-relational properties of the substance, the "marks and traces" by virtue of which the CIC is true of a given individual. The only thing unfamiliar about my claim is the idea, clearly *Leibniz's* idea, that the non-relational properties in question pertain properly to the *form* of a corporeal substance, the soul of an organism that makes up Leibniz's Aristotelian world at this time, as I argued in §III. And if the CIC resides in that way in the form of a corporeal substance, then the explanation of a property an individual has in terms of its CIC is, in essence, an explanation in terms of its form. It should be noted, though, that given the idiosyncratic conception of an individual substance that Leibniz espouses in, for example, DM8, although Leibniz allows explanations of the properties of bodies in terms of form, these explanations will not be *precisely* the explanations his Scholastic contemporaries would have given. Leibniz's substances are *individuals*, and their forms contain *all* of the properties that pertain to them as *individuals*. There is for Leibniz no form of earth or water in terms of which to explain heaviness, and no form of air or fire in terms of which to explain lightness. One can explain why George the Corporeal Substance falls toward the center of the earth in terms of *his* form (CIC), but not in terms of the forms of the Aristotelian elements of which he is composed. Furthermore, Leibniz is not committed to the position that any organism in our experience (human beings excepted) are substances, and thus have forms.[218] Consequently, Leibniz is not committed to the position that the explanation of the behavior of the animals and plants we see around us must be given in terms of their own particular forms; there may just not be any such things. But despite these qualifications, it is still fair to say that for Leibniz, at this time, one can still give explanations of *all* phenomena in the visible world which are in form, so to speak, if not in content, Scholastic explanations in terms of essence and innate tendencies.

But, at the same time, Leibniz would insist, there is *another* way of explaining what goes on in the world, in terms of the size, shape, and motion of the corpuscles that make up bodies, appealing to the laws that God has imprinted on corporeal substance. Leibniz is not explicit about just how his conception of corporeal substance commits him to the mechanists' program for physics. But the following is a plausible account of what Leibniz might have thought. Corporeal substance, that which is real in bodies, is composed of form and matter, that is, primitive active and passive forces. The modes of corporeal substance are, thus, derivative active and passive forces, and that, presumably, is all. These derivative forces give rise to extension, motion, and the resistance to motion (and, presumably, nothing more) in accordance with the laws God has imprinted on the form, the primitive active force. The connection between corporeal substance and primitive force, primitive force and derivative force, and derivative force and the modes of extension would *seem* to dictate that the only properties corporeal substances really have are derivitive forces and the motion they cause, and thus, all of the behavior of corporeal substances and that of the inanimate bodies that they go to make up can be given in terms of force, the modes of extension, and the laws that govern them.

Thus, the behavior of bodies can be explained either as the Scholastics do, in terms of forms, or as the moderns do, in terms of the laws of motion. And, Leibniz claims, these two modes of explanation will always coincide! Leibniz thus writes in a letter to Thomas Burnett from 1697:

> ...I believe that everything happens mechanically, as Democritus and Descartes desire...and that nevertheless everything also happens vitally and in accordance with final causes, everything being full of life and perceptions, contrary to the opinion of the Democriteans.[219]

And, Leibniz says even more explicitly in a letter to Des Billettes from

1696:

> I believe that everything really happens mechanically in nature, and can be explained by efficient causes, but that, at the same time, everything also takes place morally, so to speak, and can be explained by final causes. These two kingdoms, the moral one of minds and souls and the mechanical one of bodies, penetrate each other and are in perfect accord through the agency of the Author of things, who is at the same time the first efficient cause and the last end.[220]

Leibniz's metaphysical physics, thus, reconciles two apparently competing kinds of explanation, the Scholastic and the mechanistic. Ultimately, it offers Leibniz, the diplomat by temperament as much as by profession, a way of reconciling the two great scientific traditions of his day.[221]

So Leibniz's position allows him *both* Scholastic *and* mechanistic explanation of phenomena. In fact, the claim is that these two modes of explanation always coincide. Leibniz's position would thus appear to entail a kind of principle of toleration: if the mechanical philosophy is allowed, so should be the Scholastics' brand of science. But, Leibniz claims, even though we *can* explain everything in terms of the activity of souls/forms/entelechies, nevertheless, we *ought not to*:

> I agree that the considerations of these forms serves no purpose in the details of physics, and that they ought not to be used to explain particular phenomena. In this the Scholastics failed, as did the physicians of the past who imitated them, thinking that they could account for the properties of bodies by mentioning forms and qualities, without taking pains to examine the manner of their operation. This is as if one were content to say that a clock has a time-indicating property proceeding from its *form*, without inquiring wherein this property consists. This is, of course, enough for the man who buys it, if he turns over its care to someone else.[222]

It is difficult enough to see how, given his Scholastic metaphysics, Leibniz can even *admit* mechanical explanations. But how could he possibly grant them the *priority* in physics that he seems to want them to have?

Leibniz offers two reasons, one pretty good, one not so good. The not so good explanation is given, for example, in the introduction to a general treatise on natural science that Leibniz never completed, in a passage that seems to date from the early '80s:

> For whether we introduce God or an angel or a soul or whatever other incorporeal operative substance, the cause or mode of operating can always be explained in the truth we have about the things themselves. But the way in which a body operates cannot be explained distinctly unless we explain what its parts contribute. This cannot be understood, however, unless we understand their relation to each other and to the whole in a mechanical sense, that is, their figure and position, the change of this position or motion, their magnitude, their pores, and other things of this mechanical kind, for these always vary the operation.[223]

Thus, it is claimed, mechanical explanation is preferable to explanation in terms of soul or form, because to explain a body's behavior is to show how the behavior of the parts contribute to the behavior of the whole, and such explanation must, necessarily, be mechanical in nature. This, though, is not a very satisfactory justification for being a mechanist. Even granting Leibniz's claim that *only* mechanists can explain the whole through its parts (cf. the Scholastic notion of the mixture), one is inclined to see Leibniz's argument as begging the question. If explanation of the whole through its parts is mechanical explanation, then to justify his preference for the mechanical philosophy, Leibniz must show why the explanation of the whole through its *parts* is *itself* preferable to the explanation of the whole through its form.

But Leibniz gives another, philosophically stronger justification for his bias toward mechanist physics. Consider this passage from the *Appendix* of 1702:

> Hence, moreover, we understand that although that primitive force or form of substance (which, indeed, determines even the figures in matter while it produces motion) is admitted, yet, we must always proceed mechanically in explaining elastic force and other phenomena, indeed through the figures which are modifications of matter and the impetus which are modifications of form. And it is empty to fly immediately and in general to the form or primtive force in a thing when distinct and specific reasons ought to be given, just as it is empty to run back to the first substance, or God in explaining the phenomena of his creatures, unless his instruments or ends are at the same time explained in detail [*speciatim*], and the

proximate efficient causes or proper final causes given correctly, so that his power and wisdom are apparent.[224]

Leibniz's reasoning is none too clear in this passage and, at first glance, seems to add nothing to the first passage I quoted earlier. But the comparison he draws between those who explain everything in terms of form and those who explain everything in terms of God suggests something a bit different. Everything can, indeed, be explained in terms of form, just as everything can be explained in terms of God. But unless some specific content is given to these explanations, they are trivial. For example, suppose that at some given time George the Corporeal Substance moves to the left rather than to the right. One true explanation of this is that God wills that he move left; another, Leibniz claims, is that his form (CIC) determines that he move left. But these explanations are without content; *precisely* the same explanations can be given for *any* event. As Leibniz puts it in the *Specimen Dynamicum*, the bare appeal to soul or substantial form "relates only to general causes which cannot suffice to explain phenomena."[225] What we need to turn these empty (but true) statements about God and form into genuine, explanations with content, is something specific to the case at hand. For explanations in terms of God's will, the additional content must include something relating to God's ends in causing George to move left, and, perhaps, the means he used to bring about his will. And in the case of an explanation in terms of form, the additional content must involve something about the specific behavior that God programmed into that form.

Now, if we had *a priori* knowledge of the CIC that God programmed into a given form, then this would certainly suffice; knowing the specific content of George's form we could, in good conscience appeal to that content to explain his behavior. But in general, when dealing with the overwhelming majority of corporeal substances that inhabit Leibniz's world, we have no knowledge whatsoever about the *specific* properties of their forms.[226] But, on the other hand, we do know that *one* thing God programmed into each and every form is the laws of mechanism, the laws governing force and resistance, Leibniz claims. And thus, Leibniz seems to claim, an explanation in terms of the laws of motion is more informative than an explanation merely in terms of form, given our ignorance of any *other* specifics about the CIC of a corporeal substance. Leibniz's claim is not unproblematic. There is a sense in which mechanistic explanations *can* be every bit as vacuous as the Scholastic explanations they sought to replace; a mechanist can just as blithely say that an event happened because it follows from the laws of mechanism together with the (unknown) initial conditions as a Scholastic could say that the same event follows from an unknown form.[227] But, despite its weaknesses, the argument gives us something of a handle on why Leibniz thought he could say that though both Scholastic and mechanistic accounts of the same phenomenon are always available, and always both true, the mechanistic explanation is preferable. On my reading, mechanistic explanations of the phenomena are preferable not because they are *truer* than the Scholatics' account, only more informative. When understood properly, Leibniz's mechanical explanations do not *replace* Scholastic explanations, but are *grounded* in them; mechanical explanations are a schematic and partial way of describing what goes on in the form, what *all* forms have in common. It is what we have to fall back on in our ignorance of what specifically God programmed in.

And with this, my argument is concluded. I have tried to sketch out how in the 1680s and '90s Leibniz's version of the mechanical philosophy, his dynamics, was grounded directly in the notion of corporeal substance. There are, of course, many questions that I had to leave untouched. The most obvious omission concerns Leibniz's later thought. I have granted that the standard view of Leibniz and the foundations of his physics, a metaphysically real world of mind-like monads and a phenomenal world of physical bodies acting in accordance with the laws of physics, is indeed the picture of Leibniz's later years. But, one might well ask, *when* does Leibniz adopt this later view, and *why*? This and other important questions must be left unanswered for the moment at least. But what I think I have shown is that the physics of Leibniz's important middle years is *not* a science merely of the way things appear; it is not merely a science of the phenomena. Leibniz's physics is a science of corporeal substances, things that are real in the fullest sense. In that way, one can say that, at least when it first emerged, Leibniz's mature physics was intended to describe the world as it *really is*. This will change as Leibniz's thought develops. But, I would insist, it is only later in Leibniz's life, after he has put aside his serious work in physics that his science loses its grip on reality.

Department of Philosophy
University of Chicago

References

Table of Abbreviations

AT — C. Adam and P. Tannery, eds., *Oeuvres de Descartes* (Paris: 1964-1978). Citations by volume and page.

C — Couturat, Louis, ed., *Opuscules et Fragments Inédits de Leibniz* (Paris: 1903)

CA — *The Leibniz-Arnauld Correspondence.* See textual note (iii) below.

DM — *Discours de Metaphysique.* Citations by section numbers. See textual note (ii) below.

Fardella — Leibniz to Fardella, March 1690. In L.A. Foucher de Careil, ed. *Nouvelles Lettres et Opuscules Inedits de Leibniz* (Paris: 1857), pp. 317-325. See note 71 for further textual information.

G — C.I. Gerhardt, ed., *G.W. Leibniz: Die Philosophischen Schriften* (Berlin: 1875-1890). Citations by volume and page.

GM — G.I. Gerhardt, ed. *G.W. Leibniz: Mathematische Schriften* (Berlin and Halle: 1849-1855). Citations by volume and page.

Grua — Grua, G., ed. *G.W. Leibniz: Textes Inédits d'après les Manuscrits de la Bibliothèque Provinciale de Hanovre* (Paris: 1948).

L — Leroy Loemker, ed. and trans., *Gottfried Wilhelm Leibniz: Philosophical Papers and Letters*, 2nd Edition (Dordrecht: 1969).

Lang. — A.G. Langley, trans., *New Essays Concerning Human Understanding, by Gottfried Wilhelm Leibniz; Together with an Appendix of Some of His Shorter Pieces* (Chicago: 1916).

Mon. *Monadology.* Found in GVI 607-623. Citations by section number.

NE *Nouveaux Essais sur l'Entendement.* Citations given either by book and chapter number, or by page number in the text published in series VI, volume VI of *G.W. Leibniz, Sämtliche Schriften und Briefe* (Berlin: 1962).

PM G.H.R. Parkinson and Mary Morris, eds. and trans., *Leibniz: Philosophical Writings* (London: 1973).

PNG *Principes de la Nature et de la Grace, fondés en raison.* Found in GVI 598-606. Citations by section number.

R-L Geneviève (Rodis-) Lewis, ed., *Lettres de Leibniz à Arnauld d'Après un Manuscrit Inédit* (Paris: 1952).

W Wiener, Philip P., ed. and trans., *Leibniz Selections* (New York: 1951).

Textual Notes

(i) References are given in the notes. References are generally to the original language text, followed by an accessible translation. Exceptions to this practice are the references to DM, Mon., PNG, NE, and CA. Since references to DM, Mon. and PNG are given by Leibniz's section numbers, found in every translation of these works, no special citation of translation seems necessary. The best and most available translation of the NE is Peter Remnant and Jonathan Bennett, trans., *New Essays on Human Understanding* (Cambridge, England: 1981). Since Remnant and Bennett's pagination exactly corresponds to that of the text in the *Sämtliche Schriften...*, no special reference is needed. The best available translation of CA is H.T. Mason, trans., *The Leibniz-Arnauld Correspondence* (Manchester: 1967). Since Mason gives the pagination for the French text as found in GII, once again, one reference can serve for both the text and its translation. See textual note (ii) for more textual information about DM, and (iii) for more information about CA. In preparing translations for quotation in the text, I have made liberal use of the translations available. But in every case (except one; see note 223), I have checked the translation carefully against the original, and made alterations where they seem called for.

(ii) The text of the DM I usually quote from is that given in GIV 427-463. But on a few occasions I refer to earlier versions of the DM. For these texts, see Henri Lestienne, *G.W. Leibniz: Discours de Metaphysique* (Paris: 1975). Lestienne's text is a critical edition, which attempts to indicate all of the changes Leibniz made in the process of drafting the DM, insofar as this is possible given the manuscripts that now survive. The Lestienne edition is translated, with a somewhat simplified and abbreviated critical apparatus in Peter G. Lucas and Leslie Grint, *Leibniz: Discourse on Metaphysics* (Manchester: 1953). For an account of the manuscript remains see Lestienne, pp. 14-18 or Lucas and Grint, pp. xxv-xxvii. The summaries of sections do not appear in Leibniz's final text of the DM, and so are not printed in the text of the DM given in GIV. But they do appear in the principal earlier manuscript, and are included in Lestienne's edition, and when I refer to sections of the DM I mean to include them.

(iii) The principal source for the CA that I have used is the text found in GII. But I have also made use of the texts given in R-L. The text in GII is based on versions of the letters found among Leibniz's papers at Hanover. R-L, on the other hand, is based on versions of the letters sent by Leibniz to Arnauld found among Arnauld's papers. For an account of these texts, see R-L 16-18. R-L includes only some of the documents found in GII; it lacks all of Arnauld's letters to Leibniz, some of Leibniz's letters to the Landgrave Ernst von Hessen-Rheinfels, and two of Leibniz's drafts for letters to Arnauld. But in the texts that they share, there are many interesting variants. Some are of little philosophical importance, I think. For example, the term "entelechy" that appears repeatedly in the text of the letter to Arnauld, 9 October 1687, as given in GII 111-128 is given either as "form" or "substantial form" in the version that Arnauld received (R-L 78-95). But there are other, extremely important divergences between the two texts. For example, an important passage on primary and secondary matter given in GII 119-120 is missing in the letter Arnauld received (R-L 87). There are differences of opinion about the divergences between the two texts. Georges LeRoy generally assumes that Gerhardt's text is earlier, and that the text given in R-L represents last-minute alterations Leibniz made just before sending the letters to Arnauld. See, e.g., G. LeRoy, ed., *Leibniz: Discours de Metaphysique et Correspondence avec Arnauld, Introduction, Texte, et Commentaire*, 3rd ed. (Paris: 1970), p. 311, notes 24, 27, 28, 30, 31, etc. (Rodis-) Lewis, on the other hand, considers the bulk of such changes (at least those texts found in GII but not in R-L) to be later additions and changes Leibniz made probably in 1695 (but perhaps in 1707) when Leibniz considered publishing the CA. See R-L 19. In general I side with (Rodis-) Lewis. But at the same time, while Gerhardt's text may be later, it is, on many points clearer and more explicit than the texts Arnauld actually received; many of the changes seem to me to be in the way of clarifications of Leibniz's ideas, in the spirit of the original, but making explicit ideas that are only implied or suggested in the original texts, rather than changes in Leibniz's doctrine. See, on this, the discussion in note 54, e.g. Consequently, the text I quote from in the body of this paper is the perhaps later but more explicit text given in GII. When the variants are merely terminological, as they are in the case of Leibniz's later use of the term "entelechy," they will pass without remark. But when the differences between GII and R-L are significant, I shall indicate them in the notes.

Notes

1. One might also claim that for Leibniz, bodies are phenomenal in a different, almost Berkeleian sense, the vision that the world of monads share in common, a vision that does not correspond to any external reality. For this reading, see, e.g. Montgomery Furth, "Monadology," in Harry Frankfurt (ed), *Leibniz: a Collection of Critical Essays* (Garden City, New York: 1972); and John Earman, "Was Leibniz a Relationist?." *Midwest Studies in Philosophy* 4 (1979), 263-276, esp. §4.
2. Jacques Jalabert, *La Théorie Leibnizienne de la Substance* (Paris: 1947), p. 26; see also pp. 37-8, 41-2, 54-5; and Fernand Brunner, *Études sur la Signification Historique de la Philosophie de Leibniz* (Paris: 1951), pp. 216-220. Robert

M. Adams suggests something similar in a suggestive and valuable recent study. See "Phenomenalism and Corporeal Substance in Leibniz," *Midwest Studies in Philosophy* 8 (1983), 217-257, esp. p. 224.

3. See Martial Gueroult, *Leibniz: Dynamique et Mëtaphysique* (Paris: 1967), pp. 205, 207. See also Jalabert's reply to this reading in *Op.Cit.*, pp. 48f.
4. Gueroult, *Op.Cit.*, p. 205.
5. In a few pages that deserve to be better known, C.D. Broad notes an important transformation in Leibniz's thought somewhere between 1699 and 1706. See Broad, *Leibniz: An Introduction* (Cambridge, England: 1975), pp. 87-90. According to Broad, Leibniz is basically an Aristotelian in the '80s and '90s, and only later does he adopt the idealism of the *Monadology.* While I shall argue that matters are somewhat more complex than that, much of the essay that follows can be regarded as an attempt to sharpen, expand, and carefully document some of Broad's observations. Broad's thesis is also developed, in a somewhat different way in Louis Loeb, *From Descartes to Hume* (Ithaca: 1981), §§32-35. Loeb, though, sees the metaphysics of the '80s and '90s (he sets 1704 as the turning point) not as Aristotelian, but as Cartesian: "Leibniz began working within a dualist framework of minds and bodies." (p. 299). Loeb argues that this account is necessary for an understanding of Leibniz's claims about the non-interaction of substances, and for an appreciation of Leibniz's original motivation for pre-established harmony. While Loeb is correct in noting important changes in Leibniz's views, it will become clear that Loeb has misconstrued Leibniz's earlier writings in important ways.
6. I shall follow common practice in calling Leibniz's basic notion here the *individual* substance, as Leibniz often does in DM. But it should be noted that in the DM Leibniz sometimes uses the term "*substance singulier*" (DM9) or "*substance particulier*" (DM14, 15) for the same notion, suggesting that "individual substance" may not be a genuine technical term in the DM. It is also interesting to note that the term, "simple substance," so prominent in later writings, does not seem to appear at all in the DM or the CA.

7. See DM22, 23, 33.
8. See DM10-12, 34. The phrase quoted is from DM12.
9. See DM12, 21.
10. DM12, emphasis added. The something to which forms are added, while connected with extension, is not intended to be Cartesian extended substance, as I shall later show. For a different view, see Loeb, *Op. Cit.*, pp. 300-303, who *seems* to hold that in the DM and CA, Leibniz was tempted to simply distribute (Cartesian) minds through (Cartesian) bodies to deal with the problems he raises for Descartes' account of body.
11. DM34. There is a certain hesitancy throughout these passages about whether or not any bodies in the world (with the exception of those connected with human minds) are in any way substantial. For example, an earlier draft of DM34 begins (with some alteration due to a fragment of a clause Leibniz dropped):

    > I am not trying to determine if bodies are substances, speaking in metaphysical rigor, or if they are only phenomena like the rainbow, and thus if there are substances, souls, or substantial forms which are not intelligent. But supposing that bodies which make up an *unum per se*...

    The claim here, weaker than the one that appears in the final text of the DM, is that *if* bodies are to contain something substantial, then they must have forms. For similar hesitancy, see the earlier drafts of DM11, 12, and 35. In these cases, too, the hesitancy present in the hypothetical mode of presentation is missing from the final text. It is interesting that all of these passages survive into Leibniz's final ms. of DM (Lestienne's "Copie B") before they are deleted. It is also interesting to note that there is a similar hesitancy well into the CA (see GII 71, 72, 77), a hesitancy that only disappears with the letter of 30 April 1687 (GII 98). This suggests that the hesitancy was removed only in the course of dealing with Arnauld's objections, after what Leibniz originally took to be the final version of the DM. I would like to thank Robert Sleigh for emphasizing the significance of these passages to me. For more on the textual history of DM, see my textual notes (ii) on the DM appended to the table of abbreviations.

12. See GI 420 and R-L 107.
13. See, e.g. GII 74, 76-77, 97, 98, 120, 121, etc.
14. GII 96. See also GII 58, 72, 97, 118.
15. GII 118.
16. GII 72.
17. GII 97. We shall discuss what Leibniz means by phenomenal in this context below in §II.
18. GII 96.
19. See the discussion of this question in note 11.
20. GII 86-87.
21. GII 97.
22. GII 96. Later we shall discuss Leibniz's claim that extension and thus bodies cannot be made up of points. For Leibniz's argument against atomism, the claim that bodies are made up of uniform, infinitely hard small bodies, see, e.g., texts from 1690 Leibniz wrote against atomism, GVII 284-288, and the discussion of atomism in the *Specimen Dynamicum* of 1695, GMVI 248-249 (L 446-447). See also the discussion in Gueroult, *Leibniz*, pp. 98ff, and in Bertrand Russell, *A Critical Exposition of the Philosophy of Leibniz*, 2nd ed. (London: 1937), §45.
23. GII 58.
24. GII 72.
25. B. Russell, *A History of Western Philosophy* (New York: 1972), p. 583.
26. GII 86. See also GII 87, 106.
27. GII 97.
28. It should be noted that while the reply to Arnauld excludes the claim that bodies are simple aggregates of souls or forms, it doesn't exclude the possibility that all reality is mental in another sense. The reply is consistent with the claim that corporeal substances, the unities that make up bodies, are organized packets of souls, souls organized so as to form complex substances. This is the account that Adams defends in "Phenomenalism and Corporeal Substance," §3. However, one might expect that if this were Leibniz's position, he would have taken this opportunity to inform Arnauld of it. This question is taken up in more detail below in §II.
29. GII 76.

30. *Ibid.*
31. GII 75. The position is attributed to the "last Lateran Council" in this text, but clearly with Leibniz's approval.
32. GII 120.
33. GII 76.
34. See GII 73, 75.
35. See GII 120.
36. GII 77. See also GII 72-73, 75, 76-77.
37. I'm not sure what to say about the quote I earlier gave from GII 72: "...The substance of a body, if bodies have one, must be indivisible; whether it is called soul or form does not concern me." This passage *seems* pretty clearly in opposition to the reading I am proposing insofar as it *seems* to suggest that that which is real in bodies is just souls or forms. But this runs so counter to the rest of what Leibniz seems to say in the CA that I am certain that there must be a way of rendering the passage consistent with my reading of the rest of the CA. Now, when Aristotle discusses the term '*ousia*,' later translated as 'substance', he sometimes considers form to be the proper signification of that term. On this, see Léon Robin's note in André Lalande, *Vocabulaire Technique et Critique de la Philosophie*, 8th ed. (Paris: 1960), pp. 1048-49. Robin further claims that "the form and the quiddity, following the logical and ontological tendencies which dominate Aristotelianism, are substance more immediately than the individual, which...is a composite [of form and matter], while the form and quiddity are simples." Unfortunately Robin gives no references later than Aristotle. But perhaps it may be possible to fit Leibniz's statement into this tradition. On such a reading, since the form is properly identified with the term 'substance', form is, indeed, the substance of body, but, nevertheless, the unities that ground the extended bodies of everyday experience are composite entities, forms (substances) that unite bodies. It may also be possible to understand the passage in another way. Forms or souls may be the substance of a body insofar as it is by virtue of having a form or soul that a body is a genuine substance. It is in this way that Leibniz sometimes seems to understand the similar claim that "forms constitute substances;" see the

discussion of this locution below in note 135. And, finally, it is also possible that the phrase is a slip of the pen. The passage is part of a draft, and was not sent to Arnauld. And in the parallel passage of the letter he actually sent (GII 76-77), Leibniz's text can easily be interpreted consistently with the doctrine I am trying to attribute to him.

38. GII 98.
39. GII 161. This was not in the text sent to Arnauld. See R-L 93.
40. GII 118. The version of this passage given in G is somewhat richer and more explicit than the version Arnauld actually received. Cf. R-L 85-86.
41. GII 120.
42. GII 119. This is a passage not found in the letter Arnauld received. See R-L 87, note (1). Secondary matter will be discussed below in §II.
43. Given in Etienne Gilson, *Index Scholastico-Cartesien* (Paris: 1979), p. 126.
44. Fr. Eustacius a S. Paulo, *Summa Philosophiae Quadripartita*...(Cambridge, England: 1648), pp. 123-124.
45. In referring to the forms as souls, and comparing the composition of a corporeal substance with that of a human being, Leibniz seems to imply that all forms are immaterial substances. If this is, indeed, Leibniz's position, then he is departing from the orthodox position of, say, St. Thomas, for whom the only substantial forms that are substances, capable of existing separate from matter are rational souls, human or angelic. See, e.g., St. Thomas Aquinas, *On Being and Essence*, trans. by Armand Maurer (Toronto: 1968), pp. 51-59; *Summa Theologica* q 75 a 2-3. But if this was Leibniz's position, he was not the first to interpret Scholastic doctrine in this way. See, e.g., Descartes' definition of form in ATIII 502, and the discussion in Étienne Gilson, *Études sur le rôle de la pensée médiévale dans la formation du système cartésien* (Paris: 1975), pp. 162-63. It is interesting to note, though, that in a text discussed below in note 82, Leibniz seems to deny that forms, even including the human soul, are substances. And, as discussed below in §III, even if forms are substances, Leibniz held that they *never* exist in nature

without bodies. So, even if forms are substances, they never have any opportunity (short of a miracle) to display their capacity for independent existence.

46. DM11. See also DM10, GII 58.
47. GII 119. This, again, is from a passage not sent to Arnauld. See R-L 87 note (1).
48. See GII 65ff for Arnauld's initial response to Leibniz's use of the term 'substantial form' on GII 58.
49. GI 198-99 (L190).
50. It is common practice in the secondary literature that deals with Leibniz's debt to Scholastic Aristotelianism to assume that Leibniz is a monadologist throughout his mature writings. It is no wonder, with that pre-supposition, that commentators invariably find the connections between Scholastic doctrine and Leibnizian metaphysics superficial at best, even when Leibniz is quite consciously appealing to Scholastic notions of form and matter. See, e.g., Joseph Jasper, *Leibniz und die Scholastik* (Münster: 1898/99), pp 39ff; Fritz Rintelen, "Leibnizens Beziehung zur Scholastik," *Archiv für Geschichte der Philosophie* 16 (1903), pp. 157-188 and 307-333, esp. 326ff; L. Jugnet, "Essai sur les Rapports entre la Philosophie Suarézienne de la Matière et la Pensée de Leibniz," *Revue d'Histoire de la Philosophie et d'Histoire Général de la Civilisation,* Fasc. 10 (15 Avril 1935), pp. 126-136; Hervé Barreaux, "La Notion de Substance chez Aristote et Leibniz," *Studia Leibnitiana* supp. 14 (1972), 241-250.
51. For the logical argument, see DM8-11, and for the dynamical argument see DM12, 18, 21.
52. Eustacius, *Op.Cit.*, p. 119.
53. For a concise account of the Scholastic distinction, see Joseph Owens, "Matter and Predication in Aristotle," in Ernan McMullin (ed), *The Concept of Matter in Greek and Medieval Philosophy* (South Bend, Indiana: 1963), pp. 79-95, esp. p. 86.
54. GII 119. The bulk of this passage was not in the letter Arnauld received. See R-L 87 note (1). In fact, the original letters sent to Arnauld contain no *explicit* distinction between primary and secondary matter. However, I think that the distinction is implicit in those letters, and helps make clear the doctrine there, as Leibniz himself seems to have realized when

he added the bulk of this passage to the original letter. Cf. my textual note (iii) on CA appended to the table of abbreviations.

55. GII 100. Note the variants in R-L 72.
56. GII 107.
57. GII 120.
58. See, on this, A. Boehm, *Le "Vinculum Substantiale" Chez Leibniz* (Paris: 1962), pp. 35-58.
59. For an interesting account of the way the form harmonizes with its body, see R.M. Adams, "Phenomenalism and Corporeal Substance," §2.
60. These later denials are in response to an attack on Leibniz's doctrine of pre-established harmony that Boehm, *Op.Cit.* has convincingly argued is central to Leibniz's later thought on the *vinculum substantiale* in particular and to his thought on composite substance in general. The attack is found in Rene-Joseph de Tournemine, "Conjectures sur l'Union de l'Ame et du Corps" and "Suite des Conjectures...," in *Mémoires pour l'Histoire des Sciences et des Beaux Arts* (*Mémoires de Trévoux*), May 1703, pp. 864-875 and June 1703, pp. 1063-1085. Tournemine's objection is quite simple. Calling to mind Leibniz's often repeated two-clock example for illustrating pre-established harmony (see, e.g., GIV 498-500 (L459-60)), Tournemine writes: "Thus correspondence, harmony, does not bring about either union or essential connection. Whatever resemblance one might suppose between two clocks, however justly their relations might be considered perfect, one can never say that the clocks are united just because the movements correspond with perfect symmetry." (pp. 869-870) Leibniz's first reaction was to deny that there is anything to union over and above harmony. See, e.g. his remarks to DeVolder in 1706, GII 281 (L 539). But in the response he published in the *Mémoires de Trévoux* in 1708 (GVI 595-6), he suggests that union, like the mysteries of the faith, surpasses philosophical understanding. Note, by the way, that the title of the piece Leibniz published embodies a mistaken reference to the original publication of Tournemine's critique; contrary to what Leibniz implies, it was *not* published in March 1704.

61. GII 58.
62. GII 75. See also GII 112, 136.
63. For a number of attempts to explicate this notion, see McMullin, *Op.Cit.*
64. Gilson, *Index*, p. 169. See also *ibid.*, §§59, 272, 273, and Eustacius, *Op.Cit*, pp. 120, 123.
65. Gilson, *Index* p. 173. See also Eustacius, *Op.Cit.*, p. 122.
66. Eustacius, *Op.Cit.*, p. 120.
67. For an account of primary matter, understood in this way as internal to a monad, see, e.g. B. Russell, *Leibniz*, §86. It should be remarked that Russell also recognizes a different sense of primary matter, that found in the dynamics. On his view "*Materia prima*, as an element in each monad, is that whose repetition produces the *materia prima* of the Dynamics." (p. 144). What I shall attempt to establish in this section and §V below is that the primary matter of the metaphysics is the same as that of the dynamics, and in both cases, is something outside of the soul-like forms, at least in the context of the Aristotelian metaphysics that, I claim, Leibniz held in the period of the CA. It should be noted that the attribution of primary matter to an incorporeal substance, like the monad, is very much within a certain Scholastic tradition. St. Bonaventura, for example, among many others, held that angels and human souls involved both form and primary matter. See, e.g., St. Bonaventura, *In I Sent.* d 8 p 2 a 1 q 2; *In II Sent.* d 3 p 1 a 1 and d 17 a 1 q 2. Having matter in this sense, though, is something quite different from being a body or a corporeal substance for both Bonaventura and Leibniz of the monadology. That Leibniz doesn't have this notion of matter in mind in the context of what I have called the Aristotelian metaphysics is clear from the fact that he talks repeatedly of *corporeal* substances.
68. GVII 322 (L 365). This notion of primary matter is also suggested in a number of the letters to DesBosses. See, e.g., GII 324-5, 378. In one passage, GII 435 (L 600) Leibniz suggests that the primary matter of extended bodies derives from the primary matter (called there passive power for reasons that will become clear in §V below) of individual monads, as, I remarked in the previous note, Russell held. The

developments that led Leibniz to the internalization of primary matter in immaterial substance is the main theme of Georg Wernick, *Der Begriff der Materie bei Leibniz in seiner Entwicklung und in seinen historischen Beziehungen* (Jena: 1893). I would disagree, though, with the account Wernick gives of corporeal substance at the time of the CA on pp. 27-29.

69. GII 119. The reference to *extended* mass here suggests that the target *may* be Cartesian extended substance, though the last clause, referring to the Scholastic doctrine of act and potency, suggests that the doctrine of primary matter is at issue as well. For statements of the Scholastic doctrine that form gives actuality to matter see, e.g., the references cited below in note 95.
70. GII 118-119.
71. The text is published in two places, in L.A. Foucher de Careil, ed., *Nouvelles Lettres et Opuscules Inedits de Leibniz* (Paris: 1857), pp. 317-325, and in the appendix to Ludwig Stein, *Leibniz und Spinoza* (Berlin: 1890), pp. 322-325. There are some significant differences between the two transcriptions; Foucher de Careil, e.g., seems to drop a sentence on p. 322, line 10 that Stein gives, but on the other hand, includes more of the ms. than Stein gives. Both texts should be consulted. For ease of reference, though, in the notes below I shall refer to the text as simply "Fardella," and give the pagination as found in Foucher de Careil's transcription.
72. For a general discussion of Fardella and his relations with Leibniz, see Salvatore Femiano, "Uber den Briefwechsel zwischen Michelangelo Fardella und Leibniz," *Studia Leibnitiana* 14 (1982), 153-183.
73. For example, on Fardella, p.320, Leibniz seems to assert that the soul is a substance, while on p. 322 he seems to deny it. And on p. 320 he asserts that the (corporeal?) substance is not a part of a body, while on p. 322 he admits that there are senses in which substances can be considered parts of bodies. The Fardella letter, thus, seems to be more of a working paper than a polished statement of Leibniz's position.
74. The quote is from C 522 (L 270). Cf. also GII 370 (L 598) where the Latin is "*terminationes.*" Points are called

"extremities of extension" in GIV 478.

75. Points are called "*modificationes*" in GII 370 (L 598), "*modifications*" in GIV 478, and "*modalités*" in GIV 491.
76. GII 370 (L 597).
77. GII 370 (L 598).
78. For other statements of this account of points and their relation to extension, see, e.g. GI 416; GII 300; GVI 627; GVII 560; NE 152, etc. I have not emphasized chronology here since the position is one Leibniz seemed to have held from at least the early 1680s to the end of his life. Russell offers a different account of the modality of points. "They are thus mere modalities, being a mere aspect of the actual terms, which are metaphysical points or monads." (*Leibniz*, p. 105). I can see no textual support for his reading.
79. Fardella, p. 323.
80. Fardella, p. 322. It should be noted that this passage somewhat carelessly mixes two different analogies, one between points and substances, and another between points and souls. It is the latter that interests me here. That Leibniz considered them different analogies, and did not identify the souls of the second analogy with the substances of the first is shown quite clearly by the quote that follows.
81. *Ibid.*
82. To continue the analogy, one might claim that just as points are modal, so are souls. This seems to be Leibniz's motivation for a rather puzzling statement. Immediately following the last quotation, Leibniz writes: "The soul, properly and accurately speaking, is not a substance, but a substantial form, or the primitive form existing in substances, the first activity, the first faculty of acting." (Fardella, p. 322) (The language here: "*primus actus, prima facultas activa*" suggests the Aristotelian notion of entelechy; cf. *De Anima* 412a27 and its standard Latin translations.) Cf. the discussion in note 45. This sits nicely with Leibniz's claim, discussed below in §III, that there is in nature no soul that exists without a body. But it sits poorly with other passages, even in the same letter, as remarked in note 73, in which Leibniz seems to say that the soul is a genuine substance. Be that as it may, it is *extremely* interesting that at this point, in 1690, Leibniz could even

consider the possibility that the soul, the very model of the later monad, is not a substance. A similar claim, that the substantial chain is modal rather than substantial, is considered but rejected in a 1713 letter to DesBosses because it would follow from that that "bodies would be mere phenomena." See GII 481.

83. For an interesting and closely connected, but much later use of the same geometrical analogy in the letters to DesBosses, see GII 435 (L 600).
84. GII 120.
85. The two notions of the phenomenal I find in the CA are very clearly distinguished in Grua 322-323, a passage that Grua dates as 1683-1686(?). I make no claims that Leibniz's conception of the phenomenal in the later writings is exactly the same as the notions he worked with in the mid-1680s.
86. GII 97. See also Fardella, p. 320. In writing to Fardella, Leibniz considers rainbows as phenomenal in this sense, rather than in the second sense we shall consider below, as he more often considers them. Perhaps he is thinking of the rainbow as a spatially located array of color.
87. GII 119. See also GII 76.
88. GII 101.
89. Cf. the account of the rainbow Descartes gives in Discourse VIII of his *Meteorology*.
90. GII 101.
91. GII 101. I follow here the text given in R-L 73-74; the later addition of "sensible qualities" as an example of something phenomenal in this sense seems out of place. See also GII 72, 99.
92. An indication of the substantiality of things phenomenal in this sense is a bit of technical terminology Leibniz uses elsewhere. Things genuinely real are substances, of course, but things phenomenal in this sense are what he elsewhere calls *substantiata*, or semi-substances. See C 13, 438; GII 506 (L 617). "Well-founded phenomenon" might be a good term for this sense of the phenomenal. But even though Leibniz does use the term "well-founded appearance" once in the CA (GII 118), it does not seem to be a technical term, and the term "well-founded phenomenon" does not appear at all.

Furthermore, I'm not sure how exactly this notion of the phenomenal corresponds to what he later will call well-founded phenomena. While clearly they are related, I want to leave open the possibility that there might be significant differences. For a general account of the notion of the phenomenal in Leibniz, see R.M. Adams, "Phenomenalism and Corporeal Substance," esp. §§1 and 3. Adams, though, makes the assumption that Leibniz's views on these questions remain largely unchanged from the DM on (see his p. 217).

93. GII 118.
94. GII 118-119.
95. See, e.g., St. Thomas Aquinas, *Summa Theologica* I q 44 a 2 ad 3; q 45 a 4 ad 3; q 66 a 1 c; *De Principiis Naturae*, in R.P. Petrus Mandonnet, *S. Thomae Aquinatis Opuscula Omnia* (Paris: 1927), vol. I. p. 8. It should be noted, though, that this was a question of some debate among later Scholastics. On this see, e.g., Allan B. Wolter, "The Ockhamist Critique," in Ernan McMullin (ed), *Op.Cit*, pp. 124-146.
96. GII 119.
97. It is because form without matter (unlike matter without form) has unity that it *can* be considered a substance, presumably.
98. It should be noted, though, that in the CA Leibniz is careful to say that, as a matter of fact, there are no forms or souls in the world that actually exist without bodies. See, e.g., GII 124. This issue will arise again in §III in connection with the immortality of corporeal substances.
99. See DM8, 9, 14.
100. GII 116. This is not found in the letter Arnauld received. See R-L 82 note (3).
101. GII 112. This passage suggests that the mirroring thesis, the doctrine that every substance expresses the entire world of which it is a part, is completely neutral on the question of Leibniz's ontology; it is consistent both with idealism and with the sort of Aristotelianism that I have been attributing to Leibniz. The quote just given, together with the example Leibniz gives earlier on GII 112 of a perspectival projection expressing a ground plan suggests that it is even consistent with a Cartesian world of extended substances. However, the mirroring thesis and its relation to Aristotelianism and

Cartesianism is a very complex question that goes beyond the scope of this paper.

102. GII 66.
103. GII 76.
104. See GII 85, where Arnauld gives his mistaken first impression of Leibniz's position and acknowledges his mistake.
105. GII 87.
106. GII 100. See also Arnauld's similar objection from the fact that plants can be propagated from cuttings and tree limbs can be grafted (GII 85) and Leibniz's answer (GII 92).
107. It is possible, too, that neither half of the worm is animate, strictly speaking. That is, it is possible that splitting kills the worm (cf. Leibniz's theory of death below) and that the motion of both parts is purely mechanical.
108. GII 122. For Arnauld's question, see GII 108.
109. GII 123. See also GII 100.
110. Leibniz's account of generation is the exact mirror of his account of death. The natural ingenerability of souls and forms entails that generation must proceed by the transformation of a very small animal into a larger one, from an animal seed, so to speak, to animals large enough for us to see. This, in any case, is how it goes for non-rational animals. Rational souls, though, are worthy of special creation; they are custom made, as it were, for their bodies. See, e.g. GII 75, 116-117.
111. See GII 72, 75, 76, 116-117, 124.
112. See GII 116-117.
113. GII 124.
114. See, e.g., some of Leibniz's letters to Bernoulli in 1698 (GM III 551 (L 511-512)), to DeVolder in 1703 (GII 248-249, 257 (L 528, 532)), and to DesBosses in 1706 (GII 324-325).
115. This is suggested, e.g., by Leibniz's reply to Arnauld's split-worm objection on GII 100. Leibniz seems to take it for granted throughout the CA that there are no problems individuating souls or forms. To the best of my knowledge, one doesn't get an explicit discussion of that problem, nor, for that matter, a full discussion of the individuation of animals until 1703-1704, in Leibniz's elegant examination of Locke on personal identity. See NE, Book II Ch. 27. It should also be

noted that Leibniz is quite careful to set certain bounds on how an organism can transform its body. While an organism can change its body by adding or losing parts from one moment to the next, Leibniz denies the natural (vs. miraculous) possibility of metempsychosis, the transfer of a soul or form from one body at one moment to a completely different one at the next. While a soul can completely change its body over time, it must, for Leibniz, proceed part by part. See, e.g., GII 99-100, 124.

116. GII 121.

117. Insofar as the claim that all souls have bodies seems to depend on the principle of perfection, it would seem that the immortality of corporeal substances must be contingent. The whole doctrine of immortality and indivisibility of substances *seems* to rest on the immortality and indivisibility of souls. It is plausible that this doctrine, in turn, rests on the claim that souls are genuine substances. This, I think, is one of Leibniz's strongest motivations for denying that souls are unsubstantial, a position he considered in the Fardella letter. See note 82.

118. DM10.

119. GII 76.

120. DM8. In an earlier draft of DM8, Leibniz uses a different example, that of the ring of Gyges or Polycrates, claiming that while the shape of the ring is not a substance that has a CIC, the ring itself may be conceived of as having a CIC which contains all of its properties. But, Leibniz notes, this can hold only under the assumption that the ring "has a consciousness," i.e., a soul that makes it into a substance, a soul to which the CIC can attach. It is obvious why the example was dropped.

121. GII 76.

122. Mon. 62-65.

123. Mon. 66-67.

124. Mon. 1-3.

125. Mon. 17-19.

126. GII 270 (L 537). See also GIII 636 (L 659); GVI 590 (L 625), etc.

127. PNG 3. See also GII 506 (L 617); GVI 588 (L624); GVII 501-502; etc.

128. This is not to say that there is no hesitation in the DM and

CA. But the hesitation is over a different question. As I pointed out in note 11, Leibniz is not absolutely certain, at least in the beginning, whether bodies have forms or whether they are empty phenomena. But Leibniz is quite certain there that *if* something has a form, then it is a genuine substance.

129. See the accounts cited above in note 5.
130. For a discussion of what may be Leibniz's earliest flirtation with idealism in 1670 or so, see my essay, "Motion and Metaphysics in the Young Leibniz," in Michael Hooker (ed.) *Leibniz: Critical and Interpretative Essays* (Minneapolis: 1982), pp. 160-184. I should point out that the relation between this early period of Leibniz's thought and his mature work now looks considerably more complex than I represent it as being on pp. 175-176 of that essay. For further instances of Leibniz's earlier idealistic thought, see also the so-called Paris notes of 1676, I. Jagodinski, ed., *Leibnitiana Elementa Philosophiae Arcanae de Summa Rerum* (Kasan: 1913), p. 110 (L 162), and an unpublished fragment from 1680(?) quoted in Erich Hochstetter, "Von der wahren Wirklichkeit bei Leibniz," *Zeitschrift für Philosophische Forschung* 20 (1966), 421-446; esp. 439. Hochstetter also quotes two fragments, one from 1678(?), one from 1680(?) that are suggestive of the Aristotelian position that, I have argued, is found in the CA. See Hochstetter, *Op.Cit.*, pp. 429, 431. These passages are suggestive, and call for a more careful examination of Leibniz's thought before the mature works of the 1680s than I can undertake in this essay. Later in this section I shall note later instances of both Leibniz's idealism and his Aristotelianism.
131. For 1692-1693, see GIII 97; for 1696 see GVII 542 and GIV 415; for 1697 see GIII 205.
132. See GI 420. Leibniz also considered publishing the CA in 1707. See R-L 107.
133. Fardella, p. 319-320.
134. Fardella, p. 322. The Fardella letter is somewhat more reflective on certain issues than CA. The most conspicuous developments involve the conception of the soul, and the way in which corporeal substances make up bodies. On the former, as I remarked in note 82, Leibniz at least considers the possibility that souls are not substances. On the latter,

Leibniz claims that "the indivisible substance enters the composition of the body not as a part, but rather as an essential internal requisite" (p. 320). His idea seems to be that a part of a whole is of the same sort as the whole ("...geometers give the name 'part' only to those constituents which are of the same sort [*homogenea*] as the whole" (p. 323)). Thus, to use his example, a point is not part of a line, but only a shorter line is such a part (p. 320, 322). Thus, since the substance is a unity, but the body a collection, it cannot be part of a body, strictly speaking, just as one might say that a soldier is not a *part* of the army. Substances seem to be essential internal requisites of bodies in the sense that if there were no substances, then there could be no aggregates either. Leibniz wavers on this, though, and late in the letter (pp. 322-323) admits various senses in which corporeal substances and their organic bodies can be considered parts of bodies, finally conceding that "if anyone wants to call such things parts, I would allow that." (p. 323)

135. For example, Leibniz talks of "the forms which constitute substances" and "points of substance constituted by forms or souls" (GIV 479, 483 (L 454, 456)), and Leibniz claims that the substances which ground bodies are "absolutely destitute of parts." (GIV482 (L456)). But as suggestive as these passages are of the later monads, they are not decisive. The Fardella letter shows the complexity of Leibniz's thought about parts. And in a letter written three years later to John Bernoulli, which is, as I shall later show, very much within Leibniz's Aristotelian framework, Leibniz writes: "...not the flock, but the animal, not the fish pond, but the fish is one substance. Moreover, even if the body of an animal, or my organic [body] is composed, in turn, of innumerable substances, they are not parts of the animal or of me." (GMIII 537). The sense in which forms constitute substance within Leibniz's Aristotelian framework is suggested in a passage from the earlier draft of the *New System*. Leibniz writes: "Thus I found that in nature, outside of the notion of extension it is necessary to employ that of force, which renders matter capable of acting and resisting.... This is why I consider it [force] as constitutive of substance, being the principle of action, which is its

character." (GIV 472). Given the connection between (active) force and form to be discussed below in §V, Leibniz could say the same about form: it is constitutive of substance in the sense that it is the source of activity, one of the marks of substancehood for Leibniz, as he often wrote. For the connection between substantiality and activity, see e.g., the *Specimen Dynamicum* of 1695, GMVI 235 (L 435), or the "Correction of Metaphysics" of 1694, GIV 469-470 (L 433).

136. GIV 473.
137. GIV 491.
138. GIV 492.
139. GMIII 542.
140. GIII 227. The connection between form, matter, and force will be discussed below in §V.
141. GIII 260. See also a near contemporary letter to Bernoulli, GMIII 536-537 and a passage from CA, GII 119-120.
142. GIV 510 (L 503).
143. GII 171 (L 517).
144. GMIII 542. This conception of the monad explains a passage Leibniz wrote to Bernoulli in a letter from about the same time. Leibniz wrote: "How far a piece of flint must be divided in order to arrive at organic bodies and hence at monads, I do not know. But it is easy to see that our ignorance in these things does not prejudice the matter itself." (GMIII 552 (L 512)). This passage is, of course, utterly unintelligible if the term "monad" is to be understood as Leibniz uses it later in the *Monadology*. The first occurrence I know of the term "monad" (as opposed to the related term "monas") is in a letter to Fardella, 3/13 September 1696. See A. Foucher de Careil, *Op.Cit.*, p. 326. The monad is there simply defined as "a real unity."
145. GIV 511 (L 503-504).
146. GIII 454-455.
147. GIII 457-458.
148. NE 440. The monad here *seems* to be not like a soul, but like an animal *with* a soul. However, in other passages from the NE (NE 145, 231), the monad seems, like the monad of the *Monadology*, to be soul-like.
149. NE 328-329. See also NE 231-232 and 318.

150. NE 344. The distinction suggested here is even more complete than distinctions suggested earlier. It should be noted, by the way, that in Remnant and Bennett's translation, they mistakenly add an "essential" to the word "properties."
151. For the idealistic strand in 1695-1696, see the references from the "New System..." cited in note 135 and GVII 540, 542; for 1700, see GVII 551-553; for 1702 see GIII 72 and GIV 561 (L 578-579); for 1703 see GII 252 (L 530-531); for 1704-1705 see GII 270 (L 537), GIII 367, and GVII 566. I hesitate to set a final date by which I would claim that Leibniz has definitely set aside his Aristotelianism. While idealism certainly seems to dominate Leibniz's thought after 1704 or so, there are some passages written later than this that are highly suggestive of the metaphysics of the CA. See, e.g. GII 306 and GIII 657. I am also struck by certain passages in the correspondence with DesBosses in which Leibniz suggests that if there were substantial chains, then they would bind monads together and form genuine Aristotelian substances, matter, form and all. See, e.g. GII 435 (L 600) and GII 506 (L 617). One might, in fact, read the substantial chain as an attempt to capture an Aristotelian conception of substance, like the one espoused in the CA, within the context of a monadological idealism. On this, see also GVI 588-90 (L 624-25).
152. See, e.g., NE 145, 225, 378-379.
153. Russell, *Leibniz*, pp. 2-3.
154. Cf. *Ibid.*, p. 1.
155. *Ibid.*, p. v. For bibliographical information about Couturat's collection, see the table of abbreviations under 'C'. For Couturat's development of this reading, see the preface to *La Logique de Leibniz* (Paris: 1901), and, in much greater depth, "On Leibniz's Metaphysics," in Frankfurt, *Op.Cit.*, pp. 19-45.
156. See the introduction to Michel Serres, *Le Système de Leibniz* (Paris: 1968).
157. GIII 568.
158. See, e.g., GIII 205; GIII 606-607 (L 634-655); GIV 291; GIV 477ff. (L 453ff.); GMVI 240-242 (L 440-441); etc.
159. Robert Adams has demonstrated a similar complexity in Leibniz's thought on contingency. See his essay, "Leibniz's Theories of Contingency," in Michael Hooker (ed.), *Op.Cit.*,

pp. 243-283.

160. J.E. McGuire and George Gale are undertaking this project in a forthcoming paper.
161. For general accounts of the mechanist program and its history, see, e.g., Richard Westfall, *The Construction of Modern Science* (New York: 1971); Marie Boas, "The Establishment of the Mechanical Philosophy" *Osiris* X (1952), 412-541; or E.J. Dijksterhuis, *The Mechanization of the World Picture* (Oxford: 1961), esp. part IV Chapter III.
162. See Descartes, *Principia Philosophiae*, part II §§25, 27. See also Descartes' letter to Henry More, August 1649, ATV 403-404, in Anthony Kenny, *Descartes: Philosophical Letters* (Minneapolis: 1981), pp. 257-258.
163. God is given as the "universal and primary cause" of motion in *Principia* II, 36. There is good reason for believing that Descartes held that minds, both human and angelic count as genuine causes of motion too. On this see my essay, "Mind, Body, and the Laws of Nature in Descartes and Leibniz," *Midwest Studies in Philosophy* 8 (1983).
164. For the derivation of the conservation law, see *Principia* II, 36, and for the derivation of the three subsidiary laws, see *Principia* II, 37-42. It is important to Descartes that the derivation be from God's *attributes*, rather than from any *decision* He might have made to create the world in one way rather than another; there is no place for final causes in the physics Descartes wanted to create. On this, see my "Mind, Body, and the Laws of Nature...," *Loc. Cit.* On Descartes' laws of motion and the role played by God in their derivation, see Alan Gabbey, "Force and Inertia in the Seventeenth Century: Descartes and Newton," and Martial Gueroult, "The Metaphysics and Physics of Force in Descartes," both in Stephen Gaukroger (ed.), *Descartes: Philosophy, Mathematics, and Physics* (Sussex: 1980).
165. GMVI 242 (L 441)
166. The context of DM8, where Leibniz argues that individual substances have complete individual concepts and the internal marks and traces by virtue of which those concepts apply suggests that Leibniz saw this argument as intended primarily

to refute occasionalism and establish that individuals are genuine sources of activity. DM1-7 deals with God as a cause, and DM8 is presented as dealing with how "to distinguish the actions of God from those of creatures." See also passages in the CA (GII 46-47, 68-69) where Leibniz claims that it is an immediate consequence of the doctrine of DM8 that each individual substance is the source of its own activity.

167. GIV 508-509 (L502).

168. GIV 507 (L500-501). See also GIV 396-397 (Lang. 702-703). For a discussion of Leibniz's reaction against occasionalism and Spinozism, see Brunner, *Op.Cit.*, pp. 222-225. Though Leibniz rejects the Cartesian picture of God as the continual mover, he does not reject the doctrine of continual recreation on which it rests. For Leibniz's non-cartesian version of God's continual recreation, see Jalabert, *Op. Cit.*, pp. 167-178.

169. The rejection of the Cartesian picture of God as the universal shuffler of bodies happened early on in Leibniz's development and, I think, was *the* crucial move that eventually led to his mature philosophy. On this see my "Motion and Metaphysics in the Young Leibniz," *Loc. Cit.*

170. DM21. There are numerous other presentations of this same argument, mostly in the '80s and '90s. See GIV 464-465 (W100-101); GVII 447-448; GMVI 240-241 (L440-441); GVII 280-283, and in the Phoranomus of 1689, excerpts of which are published in C.I. Gerhardt, "Zu Leibniz' Dynamik," *Archiv für Geschichte der Philosophie* 1 (1888), pp. 566-581, esp. 577-581. Shorter presentations of the argument and clear references to it are found in GI 350-351, 415; GII 170 (L516-517); GII 186-187; GIII 623; GIII 636 (L659); GIV 510-511 (L503-504); GVI 320; L278. For the history and context of one presentation of the argument, see Pierre Costabel, "Contribution a l'étude de l'offensive de Leibniz contre le philosphie cartésienne en 1691-1692," *Revue Internationale de Philosophie* 20 (1966), 264-287.

171. DM21.

172. This conclusion is argued in GVII 280-281 and GMVI 240 (L440).

173. Leibniz appeals to experience in rejecting the geometrical laws in, e.g., GIV 465 (W101) and GIII 636 (L659). But, as Leibniz

knew perfectly well, a suitable physical hypothesis can render virtually any laws of motion consistent with experience. This, in fact, is Leibniz's strategy in his own early physics, in which he proposed laws of motion very similar to the "geometrical" laws he is rejecting in the argument under discussion (the laws are proposed in his *Theoria Motus Abstracti* of 1671) and rendered them consistent with experience by way of a physical hypothesis (the *Hypothesis Physica Nova*). For the classic account of this early system, see A. Hannequin, "La Premier Philosophie de Leibniz" in his *Études d'Histoire de la Philosophie* (Paris: 1908), v. 2. The better suggestion is that the geometrical laws are inconsistent with metaphysical first principles. See, e.g., GII 170 (L517); GMVI 240 (L440-441). The most obvious principle violated is the principle of equality of cause and effect, the princile that underlies Leibniz's famous conservation of mv^2 law. If the geometrical laws were right, then any collision of a moving body with a body at rest would result in an increase in mv^2, and any head-on collision would result in a decrease.

174. It should be remarked that this argument works *only* if we assume that the laws that govern the behavior of bodies in motion derive from something *intrinsic* to body. Descartes could claim that that which determines the behavior of bodies is God and His activity is such as to block Leibniz's derivation.

175. GMVI 241-242 (L441). Cf., e.g., DM18; GIV 465-466 (W101); GVII 283; Phoranomus in Gerhardt, "Zu Leibniz' Dynamik," *Loc. Cit.*, p. 580. It should be noted that what Leibniz calls force here is sometimes called power (*potentia* or *puissance*). See, e.g., GIV 395 (Lang. 701) and GIV 523.

176. GMVI 238 (L438).

177. Force is also connected with mass, of course.

178. GMVI 237 (L437). See also GIV 395 (Lang. 701).

179. This is not to say that Leibniz's conception of passive force is identical to Newton's conception of inertia. On this, see Howard Bernstein, "Passivity and Inertia in Leibniz's *Dynamics*," *Studia Leibnitiana* 13 (1981), 97-113. In what follows, though, I shall continue to use the term inertia to refer to this aspect of Leibniz's passive force.

180. The "Brief Demonstration" is given with a later appendix in

GMVI 117-123 (L296-302). For other presentations of the argument in works of the same period, see DM17, GIV 370-372 (L393-395); GMVI 243-246 (L442-444); GMVI 287-292; etc. Strictly speaking, the argument is carried out in terms of the Cartesian notion of quantity in motion, size times speed, and the point is that force is proportional to mv^2 and that this must be something different than quantity of motion. For discussions of the argument, see Carolyn Iltis, "Leibniz and the *Vis Viva* Controversy," *Isis* 62 (1971), 21-35; George Gale, "Leibniz' Dynamical Metaphysics and the Origins of the *Vis Viva* Controversy," *Systematics* 11 (1973), 181-207; and Gueroult, *Leibniz*, pp. 28-34. The main argument in the "Brief Demonstration" involves an *a posteriori* assumption, Galileo's law of free fall. For a discussion of an *a priori* argument that Leibniz offers to the same conclusion, see Gueroult, *Leibniz*, chapter V.

181. DM18. See also GII 98 and Leibniz's remarks on Descartes' definition of motion, GIV 369 (L393). Leibniz's position is an interesting one. One the one hand, he believes that there is, in a sense, a real fact of the matter about what is in motion and what is at rest; in a given collision the body with active force is, in a proper sense, in motion, and the body with passive force (resistance) at rest (though not completely, since its parts will still be in motion). But, at the same time, Leibniz is a radical relativist in physics, rejecting Newton's bucket experiment and arguing that the laws of motion must be invariant under *all* transformations. This was a position especially emphasized in the *Dynamica* of the late 1680s and some writings from 1694-1695. See GMVI 484f, 507f; GMVI 253 (L449-450); GMII 184-185; 199 (L418, 419); C590. For a discussion of Leibniz's relativism, see, e.g., John Earman, "Leibnizian Space-Times and Leibnizian Algebras," in Butts and Hintikka (eds.), *Historical and Philosophical Dimensions of Logic, Methodology, and Philosophy of Science* (Dordrecht: 1977), and Howard Stein, "Some Philosophical Pre-History of General Relativity," *Minnesota Studies in the Philosophy of Science*, 8 (Minneapolis: 1977). While this looks inconsistent, it is not. The claim is that the laws of physics that are framed

in terms of size, motion and, in general, the modes of extension, are radically relativistic, but that force, which goes beyond motion and extension, fixes what might naturally be called real motion and rest.

182. GII 120. See also GIV 394 (Lang. 700).

183. I have emphasized force as a cause of motion and resistance. But in many ways, I find Leibniz's doctrine quite unclear. It is not clear at all just how it is that force, something that Leibniz considers to be quite different from motion, is supposed to cause motion or the resistance to motion. Furthermore, Leibniz holds that force is also that which is real in motion and extension, which, for Leibniz, are phenomenal in a sense discussed above in §II. See, e.g., GMVI 235 (L436). Leibniz, in fact, doesn't seem to distinguish the two roles force plays: "...even though force is something real and absolute, motion belongs to the class of relative phenomena, and truth is found not so much in phenomena as in their causes." (GMVI 248 (L446)). The claim is obscure and its connection with the claim that force is the cause of motion is difficult. The problem is somewhat illuminated by a comparison that Leibniz draws between his conception of the relation between force and motion or extension, and the mechanist conception of the relation between motion and extension on the one hand, and color, taste, sound, etc. on the other. See, e.g., DM12; GI 392; GII 119. The suggestion seems to be that just as a certain configuration of shape, size, motion is, for the mechanist, what causes a body to be red, say, and at the same time, is that which is real in the redness of the body, so a certain distribution of forces is both the cause of the modes of extension and that which is real in them. But the status of motion and extension within Leibniz's Aristotelian metaphysics is a difficult question and raises issues too complex to enter into in this paper.

184. GIV 469 (L433).

185. GII 97-98; emphasis added.

186. GIV 473. See also GMVI 241-242 (L441).

187. GMVI 237 (L437).

188. *Ibid.*

189. GMVI 236 (L436). The passage quoted deals specifically with

derivative active force.

190. GIV 473; emphasis added.

191. GMIII 552 (L512). The same analogy is used in a number of other places, e.g., GII 270 (L537); GIII 457. Derivative force is called a modification in GII 251 (L530); GII 262 (L533); GIII 356. It is not clear in some of these passages, though, just what derivative forces are modifications of, whether they modify substances (or matter or form, the constituents of substances), or whether they are modes of aggregates of substances. On this see note 213 below.

192. GIV 395 (Lang. 701); emphasis added.

193. Cf. Lady Masham's complaint to Leibniz that "force...cannot be the essence of any Substance" (GIII 350) and Leibniz's answer: "This is doubtless because you talk of changeable forces, which one commonly conceives. But by *primitive force* I understand the principle of action, of which the changeable forces are only modifications." (GIII 356).

194. Leibniz also suggests that "primitive force is the law of the series, as it were, while derivative force is the determinate value which distinguishes some term in the series." (GII 262 (L533)). The mathematical analogy is interesting, and we shall return to it again. But I think it is not as illuminating with respect to the metaphysical status of derivative forces as their conception as modifications of form and matter. George Gale emphasizes the mathematical analogy in his essay, "The Physical Theory of Leibniz," *Studia Leibnitiana* 2 (1970), pp. 114-127, esp. p. 122f.

195. GMVI 236-237 (L436-437). See also GII 171 (L517); GIV 395-399 (Lang. 701-704); GIV 510-511 (L503-504).

196. GIV 395 (Lang. 701).

197. My account is not unlike the one given by George Gale in "The Physical Theory of Leibniz," *Loc. Cit.* Gale identifies three levels, the metaphysical level (monads), the explanitory level (corporeal substances), and the observable level (bodies). See his diagram on p. 116. My claim is that what concerns Leibniz in his physical writings in the 1680s and 1690s is Gale's explanatory and observable levels.

198. GIII 54-55 (L353). It is important to note here that Leibniz, unlike Descartes, sees God as a deliberative agent, and is quite

happy to admit final causes in physics. On this see my essay, "Mind, Body, and the Laws of Nature...," *Loc. Cit.*

199. GIV 507 (L500).
200. DM18.
201. GIV 399 (Lang. 705).
202. GIV 507 (L501). See also GIV 562 (L579); GVII 283; GMIII 545; GMVI 241-242 (L440-441).
203. See GMVI 241 (L440-441).
204. GMVI 237 (L437).
205. The most systematic such attempt occurs in the *Dynamica* from the late 1680s, where the principle of equality of cause and effect is laid down as an axiom, and the laws of motion and impact derived from it. See GMVI 435ff. The principle of equality of cause and effect is given on p. 437.
206. GII 40.
207. See note 194.
208. This is what Leibniz calls directive or progressive force. See GMVI 238-239 (L439); GMVI 495-496. In addition, the composite body will have what he calls "respective or proper force..., that by which the bodies included in an aggregate can interact upon each other." (GMVI 239 (L439)). It should be noted that Loemker's translation of this passage from the *Specimen Dynamicum* is misleading; he renders the single Latin word '*respectiva*' both as 'relative' and as 'respective.'
209. See GMVI 253-254 (L450); GMVI 496ff.
210. This is, in essence, an argument we examined earlier. See the discussion in note 170.
211. See the discussion of the phenomenality of space in §II and the discussion in note 183.
212. GII 275. See also GII 251 (L530). I take the simple substances here to be monads.
213. GII 306. See also a letter to Nicholas Remond from 1715 (GIII 636 (L659)) where Leibniz relates this conception of derivative force to inertia. Two remarks are in order here. Though Leibniz clearly holds that derivative forces pertain to aggregates of monads or simple substances, he still persists in the view that, as opposed to primitive forces, derivative forces are modal or accidental. See, e.g., GII 257 (L532); GII 270 (L537), etc. Leibniz is certainly entitled to this, as long as they

are understood to be modes or accidents, not of individual substances, but of *aggregates* of individual substances. This is, indeed, exactly how they are construed in a chart that Gerhardt gives as an appendix to a letter to DesBosses from August 1715 (see GII 506 (L617)). There derivative forces are categorized as either modifications of composite substances (genuine substances formed from a multitude of modads) or as what he calls "semiaccidents" that pertain to "semisubstances," non-substantial aggregates of monads. This leads to a second remark. Though derivative forces clearly pertain to collections of monads, collections of the basic constituents, Leibniz doesn't *always* consider derivative forces phenomenal. In the DesBosses correspondence, where Leibniz is trying to specify the conditions under which monads can come together and form a genuine composite substance, he seems to hold that while derivative forces pertain to the composite (rather than to the simple substances or monads that make up that composite), insofar as the composites can, under appropriate conditions (with the help of a substantial chain) form a genuine composite substance, derivative forces can be real, the real modes of composite substances. See, on this, the chart to which I referred earlier in this note. In this way one might read the DesBosses letters not only as an attempt to ground the theological doctrine of transubstantiation, but as an attempt to save the reality of physics. For a similar reading of the substantial chain in the DesBosses letters, see Christiane Fremont, *L'Être et la Relation* (Paris: 1981). Fremont, though, does not seem to be aware of Leibniz's conception of corporeal substance in his earlier writings. For an account of the *vinculum substantiale* in the DesBosses correspondence emphasizing Leibniz's theological motivation, see R.M. Adams, "Phenomenalism and Corporeal Substance," §5.

214. "There are individual and particular behaviors [*functiones*] appropriate to each individual natural thing, as reasoning is to human beings, neighing to horses, heating to fire, and so on. But these behaviors do not arise from matter which...has no power to bring anything about. Thus they must arise from the substantial form." (Conimbrian commentaries on

Aristotle's *Physics*, in Gilson, *Index...*, p. 127.) For Descartes' version of a Scholastic account of gravity, see ATII 223.

215. AT XI 7.
216. GII 78.
217. DM18. See also, e.g., DM10, GIV 470 (L 433); GIV 478 (L 454); L 289; C 341-342; etc. Note that, as discussed earlier, this claim rests ultimately on two claims discussed above in §5, the rejection of the occasionalist conception of God as the universal shuffler of bodies, and the rejection of the laws that, Leibniz claims, would follow from the nature of extended bodies as such.
218. See GII 72-73.
219. GIII 217.
220. GVII 451 (L472).
221. See also L289; GI 199 (L 190); GIV 559-560 (L 577-578); etc. This is close to at least some statements that Leibniz gives of pre-established harmony, at least one of whose consequences is that the behavior of an organism can be explained either in terms of the activity of the soul, or in terms of the physical laws that govern the body. Cf. the account of pre-established harmony given in my essay, "Mind, Body, and the Laws of Nature...," *Loc.Cit*.
222. DM10. See also GII 58, GII 78; GIV 345-346; GIV 397-398 (Lang. 703); GIV 479 (L454); GVII 317 (PM85); GMVI 242-243 (L 441-442); etc.
223. L 288. I have not been able to check the original Latin of this passage which, to the best of my knowledge, is available outside of the manuscripts only in Loemker's English translation.
224. GIV 397-398 (Lang. 703). See also L 289; GMVI 242 (L 441).
225. GMVI 236 (L 436).
226. The situation may be different when we are dealing with rational souls, where reasons for action may sometimes be accessible, as Leibniz suggests in one undated fragment. See C329 (W 89).
227. This idea was suggested to me by Alan Gabbey in conversation and in an unpublished lecture.

Ian Hacking

WHY MOTION IS ONLY A WELL-FOUNDED PHENOMENON

The Coriolis effect is...negligible with respect to fluvial processes, however, and deflection of the paths of rivers is not related to it. *Encyclopedia Brittanica*, Chicago, 1976. *Micropedia* Vol. III, p. 152, entry: "Coriolis Effect."

The Coriolis effect has great significance in astro-physics and stellar dynamics [etc.]...The Coriolis force figures prominently in studies...of the hydro-sphere, in which it affects the rotation of whirlpools, meander formations in rivers and streams, and the oceanic currents. *Ibid.*, Next entry: "Coriolis Force."

The Coriolis force is an inertial force first described by G.G. Coriolis in 1835. It is associated with bodies moving on rotating surfaces, so it is important to major motions on the surface of the earth. The computations involved are entirely classical and could easily have been made by contemporaries of Leibniz. Yet even now classical mechanics is not fully satisfactory: two successive entries in the latest *Encyclopedia* flatly contradict each other as to whether the Coriolis effect is significant in the study of the flow of rivers and streams.

What is more remarkable about the Coriolis force, however, is that (a) it does not exist, and (b) "it is a controlling factor in the directions of rotations of sunspots." The Coriolis force plays a major role in meteorology, for it causes big winds to blow nearly parallel to lines of

K. Okruhlik and J. R. Brown (eds.), The Natural Philosophy of Leibniz, 131–150.

equal barometric pressure, instead of rushing headlong to the centre of low pressure. All rocketry compensates for the Coriolis force, and it determines the direction of rotation of water draining out of the bathtub.

Why then do I assert that the force does not exist? Because elementary physics texts say so. The force is an artifact of the choice of reference frame. It literally does not occur in the favoured Newtonian frame of reference. It is one of many forces called a fictional or pseudoforce. For ease of calculation in meteorology or rocketry alike we commonly adopt frames of reference in which these pseudoforces occur and enter into all our calculations. Yet they are only "well-founded phenomena." I shall return to the Coriolis force below, but I thought it well to begin with the reminder that some well-founded phenomena, which are not real, are not the preserve of Leibniz's more arcane metaphysics. They are a standard part of applied Newtonian mechanics.

Motion Before Space and Time

Leibniz's thesis is of course more radical: space and time are merely well-founded phenomena. So is motion. In our days, philosophers have mostly discussed space and time, thinking that the claim about motion is only a corollary. I think that we have not heeded Leibniz. We should reason in the reverse direction. If motion is only a phenomenon in an active universe, then space and time themselves cannot be real. We must, then, see why motion is only a phenomenon. That leads us not to the empty contemplation of space and time, but to the rich and practical world of kinematics and dynamics. Once we have understood the difference between real and phenomenal quantities in mechanics, space and time are a piece of cake. So in this paper I propose a criterion for reality in mechanics. I think it is Leibniz's own. It is not so very far removed from the present basis for distinguishing real from fictional forces.

Phenomena

There are now two distinct uses of the word "phenomenon." One is peculiar to philosophy. In this usage phenomena are private, peculiar to the person perceiving, and may for example be sense-data. These are the phenomena of the philosophical doctrines called phenomenalism and phenomenology. Although Kant in no way originated this usage, he certainly stabilized it and made it a permanent part of our philosophical lexicon.

In the other usage a phenomenon is entirely public, and is either something noteworthy, or else a regularity of nature, or both. Thus, like Leibniz's contemporaries, we may speak of the phenomena of magnetism. In some branches of contemporary experimental physics one kind of specialist is known as a phenomenologist. In solid state and high energy physics people speak of phenomenology, i.e. the science of fairly low level experimental results of the sort that ought to be encompassed in high level theory. At a more commonplace level a standard college book of mechanics will say, "We present here a phenomenological approach to the evaluation of the Reynolds number" (A constant characterizing the flow of viscous liquids.)[1]

Both these senses of the word "phenomenon" and its cognates carry the implication that the phenomena are not the last word. Thus corresponding to phenomena, in the sense of sense-data, we imagine that there is some reality of the external world. Corresponding to the phenomena of electricity and magnetism we suppose that there is some underlying theory which conveys a deeper explanatory truth about this domain. There is a deeper derivation of the Reynolds number than the phenomenological one. In each case we hark back to the ancients with their philosophical contrast between phenomena and noumena, and also to Renaissance astronomy. For the stargazers, the phenomena were the positions that the heavenly bodies appeared to take up, although their real paths might be different from their apparent ones. All these phenomena of science past and present are public, and have nothing to do with sense-data.

Likewise, Leibniz primarily uses the word "phenomenon" to denote something public. There is, however, a difficulty when writing of an idealist. Consider the case of Berkeley. He is often called a phenomenalist. He himself uses a better label: immaterialist. In his own work he uses the word "phenomenon" to denote what is public, and he often himself cites the phenomena of the magnet. Since he is also an idealist and holds that everything is ultimately mental, it would be possible, although misleading, to say this: for Berkeley all phenomena, in the public sense, are also only phenomena, in the private sense. We have something of the same problem with Leibniz, but I shall persist in using the word "phenomenon" only in the public sense of something that is accessible to all of us. To say that space and time and motion are phenomena is not to say that they are private mental constructs, but to say that they are public constructs which will not occur in a deeper description of reality.

The Problem

Our task in reading Leibniz is to understand why he thinks that some public quantities are phenomenal while others are real. Which, among the quantities of physics, are real for Leibniz? Leibniz is by no means unequivocal, but the prime candidate is *vis viva.* Nowadays we say that *vis viva* is kinetic energy, but let us keep to the old term. Since about 1860 the concept of kinetic energy has been embedded in a network of concepts unknown to Leibniz. Histories sometimes imply that Leibniz sorted out the distinction between momentum (mv) and kinetic energy (mv^2) and that although it took the physics community a while to absorb the lesson, Leibniz had done the work. I was struck in reading 19th century physics how little work had in fact been done, and how people had to wrestle with *lebendiges Kraft* and connect it with other kinds and sources of energy. Much of what we might glibly attribute to Leinbiz obtained its modern sense only in the great days of thermodynamics.

Let us not then attribute to *vis viva* anything beyond the words of Leibniz himself. We have an added benefit in keeping to the old terminology, for Leibniz explicitly contrasts his living force with dead force, giving as examples of the latter centrifugal force, centripetal force and gravity.[2] Living force, he tells us, is ordinary force combined with motion. A dead force is a quantity which "solicits" a body to accelerate, whereas a living force is associated with the actual motion of a body. This is not measured by the Cartesian "quantity of motion" that we now call momentum, mv, but rather with what we now call kinetic energy, mv^2. His criticism of Cartesian physics rests largely on excellent thought experiments to show that the Cartesian emphasis on mv is inadequate to describe the interactions of colliding bodies.

We can, however, understand the critique of Descartes without yet seeing why *vis viva* is a real quantity while motion itself is only phenomenal. My task in this paper is to provide a criterion which explicates Leibniz's doctrine.

Centrifugal force

Tie a pound weight on a string and spin it around. You feel the weight pulling on your hand. We call this centrifugal force. We may even compute it, yet at the same time our teacher tells us this is not real force, but only a fictitious force, or pseudoforce. The true description of the spinning weight is as follows. Once it is in orbit, it is, at any moment, travelling in a straight line, the tangent to the circle of orbit. Were it not for the string, it would continue to travel in that straight line. However it continuously changes its direction of flight in order to stay in the orbit dictated by the string. Thus at any instant, as it changes its direction of motion, it is also accelerating towards the center of the orbit, namely your hand. This acceleration requires force, a pull towards the centre, which is called the centripetal force. You exert that force. You do not do so because the object is pulling away from you thanks to a force of its own. It is inert. The only force exerted in this system is that pulling the weight to the centre of the circle. It certainly feels to you as if the string is pulling on your hand, but, says the physics teacher, there is no such force.

Such pseudoforces abound. We all know the sensation in the pit of the stomach as the elevator accelerates. When the motor car stops suddenly, or is subject to terrible acceleration as when hit from behind, our head is pulled violently, sometimes violently enough to cause the injury called whiplash. It is just as if the head and neck were subject to some unexpected force, and indeed we speak of forces of acceleration and deceleration. But there is no such force. We are experiencing only inertia. When the car is hit from behind, the head is at rest. It carries on at rest, for if the change in speed is too instantaneous, we are powerless to use our muscles, and the head is whipped back, or rather, our bodies are accelerated while the head is not.

In the Newtonian theory of the text book, centrifugal force is a fiction but centripetal force is real. I am not entirely clear about the reference of the following sentence by Leibniz:

> I do not mean that these mathematical entities are really found in nature as such but merely that they are means for making accurate calculations of an abstract mental kind.[3]

He wrote this in the context of an example of centrifugal force. Certainly Leibniz's words might be used by a high school teacher today to describe centrifugal force. However Leibniz will say that centrifugal, centripetal and gravitational force are all in the domain of phenomena, not reality, while *vis viva* is real enough. Thus his distinction, as befits an anti-Newtonian, does not stop short at centrifugal force.

The Coriolis Force[4]

Suppose you are standing on a merry-go-round that has acquired a steady speed of rotation. I am standing on the ground watching you. I will observe only your acceleration towards the centre of the rotating platform, the centripetal acceleration. But if you are standing say by the first row of horses, and walk outwards towards the outer row of horses, I will see you undergoing an acceleration called the Coriolis acceleration. This is because although the angular velocity remains the same, (so many revolutions per minute) your linear velocity increases, since the outer horses are going faster than the inner horses. But if there is acceleration, Newton teaches that there must be a force as well. There is indeed. There is a force, which we attribute to the friction between the soles of your feet and the surface of the merry-go-round, a force that can be traced back to the motor. The engine has to work a tiny bit harder as you walk out towards the perimeter. You will feel this force, and lean into it to keep your balance. (This is *not* the Coriolis force).

I was watching you on the merry-go-round. Imagine now a third party, who is sitting still on the merry-go-round watching you. From the third party point of view, as you walk towards the outer ring of horses, you are in equilibrium. I, on the ground, see you going faster in a direction at a tangent to the merry-go-round, but relative to the observer on the merry-go-round there is no such acceleration.

How is the observer on the merry-go-round to account for the frictional force on the soles of your feet since from his point of view you are not accelerating? I shall quote the answer from an excellent elementary text. Note the nervous use of the word "real."

> ...the floor is exerting a (very real) frictional force on the man's feet. This force has one component that points radially inward and one that points...in the direction of rotation.
>
> From the point of view of the ground observer these forces are understandable and quite necessary. One is associated with the centripetal acceleration and the other with the Coriolis acceleration. The observer on the merry-go-round does not see either of these accelerations however; to him the walking man is in equilibrium. How can this be, in view of the frictional forces

that act on the soles of the walking man's shoes? The man himself is well aware of these forces; if he did not lean to compensate for their turning effect, they would knock him off his feet!

The observer on the merry-go-round saves the situation by declaring that two inertial forces act on the walking man, just cancelling the (real) frictional forces. One of these inertial forces, called the *centrifugal force*...acts radially outward. the other, called the *Coriolis force*...acts *opposite* to the direction of rotation. By introducing these forces, which seem quite 'real' to him, although he cannot point to any body in the environment that is causing them, the observer in the rotating (noninertial) reference frame can apply classical mechanics in the usual way. The ground observer, who is in an inertial frame, cannot detect these inertial forces. Indeed there is no need for them -- and no room for them -- in his applications of classical mechanics.[5]

Coriolis forces are pretty important to us, because we are on a merry-go-round, namely the planet Earth. The natural way for us to explain the motions of the winds, the direction of oceanic currents, and corrections in rocketry is in terms of these forces. But in another inertial frame of reference, these forces literally do not exist.

Note that the forces are propounded partly because they help us explain phenomena, and partly to enable us to keep the laws of classical mechanics. An alternative would be to use different laws for different frames of reference. I shall now illustrate this.

Frames of Reference and Laws of Nature

In his famous *Lectures on Physics* Richard Feynman considered two people, Joe and Moe, who use different coordinate systems. They coincide in the y and z directions. In the third direction Joe measures distances by x while Moe uses x' where $x = x' + s$. Joe measures the velocity of a moving particle by $dx/dt = dx'/dt + ds/dt$. When s is constant, Joe and Moe assign the same velocity to objects, for ds/dt is 0. Suppose that Moe's coordinate system is moving away from Joe's, at a constant velocity. Then Moe will assign a different velocity to a moving particle than Joe will. However the law of motion has force equal to mass times the acceleration, d^2x/dt^2. With constant velocity *s* the second derivative described by Moe and Joe will be the same, so once again they will employ the same law of motion. Suppose however that Moe's system is moving away from Joe's at ever increasing velocity, with acceleration *a*. Then Joe's law of motion will be:

$$(1)\ F_x = m\, d^2x/dt^2$$

Moe's, however, will be:

$$(2)\ F_{x'} = m\, d^2x'/dt^2 + ma$$

As Feynman puts it,

> That is, since Moe's coordinate system is accelerating with respect to Joe's, the extra term *ma* comes in, and Moe will have to correct his forces by that amount in order to get Newton's laws to work. In other words, here is an apparent new force of unknown origin which arises, of course, because Moe has the wrong coordinate system. This is an example of a pseudo force.[6]

Actually we can describe the situation in two distinct ways. Moe is postulating a new force, so that he can say that Newton's law is obeyed -- force is the product of mass and acceleration. Alternatively we can say that Moe is proposing a variant on Newton's laws, written down as (2) above. Much the same may be said for centrifugal force and other forces of rotary motion such as the Coriolis force. After mentioning rotation and centrifugal force, Feynman himself goes on to remind us that, *"These forces are due merely to the fact that the observer does not have Newton's coordinate system, which is the simplest coordinate system."*

Simplicity

Leibniz had great respect for simplicity. He thought that God would try to combine maximum simplicity of laws of nature with maximum variety of phenomena. He certainly encouraged us to find simple laws in nature. But he was, I think, never under the illusion that simplicity itself can be a reason for thinking that one law rather than another is absolutely true. In particular, if two laws conform to all the known phenomena, but one is simpler than the other, we can surely want to use the simpler one in our calculations as if it were true. *But that is not a reason for thinking that it is in fact the truth.*

Joe has what Feynman cheerfully calls Newton's coordinate system while Moe has a system that gives rise to a more complex law of nature. Which is the true law, the law of Joe and Newton, or the law written down by Moe? The best law to use is of course Newton's. But is it the true one? No, it is only the simplest. Both are equally good phenomenal laws. Both fit the facts perfectly. The careless might say that both are true. The stricter Leibniz would remind us that they cannot both be true, for they are incompatible. One says that a certain force is acting, while the other says that a different force is acting. Neither is absolutely true of reality. Both are merely phenomenal.

We can of course compute as if the simplest hypothesis is true. Leibniz says exactly this. Writing to Huygens in 1694, he says:

> I hold, then, that all hypotheses about the motion of bodies in different reference frames are equivalent and that when I assign certain motions to certain bodies, I do not have, and cannot have, any other reason but the simplicity of the hypotheses which I choose, for I believe that one can hold the simplest hypothesis (other things being equal) for the true one.[7]

He is not saying the simplest *is* true, but that it can be treated as if it were true.

Ptolemy-Copernicus

Let us take a brief but instructive step aside to consider Leibniz's opinion of the Copernican and Ptolemaic schemes. I speak of schemes rather than theories, for it was well known that state-of-the-art Copernican astronomy was more a programme for studying the heavenly bodies than the definitive description of their movements. Likewise the Ptolemaic scheme was a family of programmes for adapting an overall conception to the messy data of observation.

Leibniz was perhaps the first major figure to state an idea that later became both commonplace and also not taken very seriously. The planets can be correctly described in both Copernican and Ptolemaic ways. The two modes of description are what Leibniz called equipollent, and each can square with all the phenomena. Hence *neither* expresses the truth of the matter. One description seems to be immeasureably simpler than the other. One leads the mind to deep propositions about nature. Leibniz himself hit upon the inverse square law before Newton published. Such a law is invited by a heliocentric model, but not by any story of epicycles. Leibniz thought Copernican theory the right one to use, and in this he did not differ from astronomers in holy orders. But according to Leibniz, our system of heavenly bodies is not, in reality, either heliocentric or earth-centered. Leibniz pinned one of his vain hopes of reconciliation on this judgment. If only he could get the Vatican to see the point, the rejection of Galileo could be withdrawn.[8]

When we debate the merits of Copernican and Ptolemaic schemes we are at the level of phenomena, not reality. This opinion is important to Leibniz's philosophy. It is based upon an implicit but fundamental principle. Let there be two descriptions of the same phenomena. We say that the two descriptions are both *empirically adequate* if they both fit the phenomena. We say that they are *literally inconsistent* if they could not both be literally true. Thus, "the sun is stationary, and the earth moves in orbit about it" is literally inconsistent with "the earth is stationary, and the sun moves in orbit about it." Leibniz's principle is, that if the two descriptions are empirically adquate but literally inconsistent, then neither of them can be held to express the real truth of the matter. So far as the evidence goes they must alike be regarded as phenomenal descriptions, no matter how well founded. Neither is a description of reality.

When Leibniz speaks of two hypotheses being equipollent, he usually has in mind that both are (or can be made) empirically adequate, while they are literally inconsistent.

Invariant Laws of Nature

The connection with Feynman's accelerating reference frame is obvious. Joe describes a particle using frame of reference R and law of nature L, while Moe uses frame of reference R^* and law of nature L^*. Joe and Moe are both empirically adequate but literally inconsistent with each other. We Newtonians prefer Joe to Moe. Leibniz says we do so on grounds of simplicity, not truth.

It may seem odd to speak of the literal truth of R, a frame of reference. That is because the very term "frame of reference" has by now some relativistic overtones. But if we take R literally, it is saying that certain points are at rest while others are in motion. R^* says something inconsistent with that. The analogy with the rival planetary schemes is clear.

Unless we have a canonical frame of reference we cannot in general ask whether L is a law of motion. We can ask only whether $L\&R$ together provide an empirically adequate law. According to the principle that I attribute to Leibniz, neither the L nor the L^* in Feynman's example can be the real truth of the matter. $L\&R$ and $L^*\&R^*$ are equipollent. Simplicity may persuade us to use one over the other, but it is not a ground for thinking that one of them is the truth.

Yet not all laws of mechanics are sensitive to frames of reference. *Vis viva* is conserved in interactions between colliding objects. *This conservation law holds in any frame of reference.* Of course the amount of *vis viva* attributed to a body will depend on its velocity and hence on the frame of reference. But it does not matter whether we have a frame describing A as at rest and B in motion towards it, or A moving slowly North and B rushing North to catch up to A...no matter what frame of reference, we can compute the consequence of a collision by applying the straightforward law of conservation to the *vis viva* assigned in that frame of reference. In every case the calculated outcome will be empirically adequate. The conservation law is *invariant* with respect to change of frame of reference even though velocities are not. We do not have (incompatible) equipollent laws suitable for different frames. We have one law that fits all frames. Incompatible members of a set of equipollent laws are not candidates for the truth because they cannot all be literally true. *Empirically adequate invariant laws are candidates for the truth.* The $F = ma$ is one of infinitely many equipollent laws which are well founded in phenomena but which cannot be held to be real. The law of conservation of *vis viva* might in contrast be a real law of nature.

Quantity and Law

Like all idealists Leibniz is something of a holist. Quantities are not to be understood as if they had an existence on their own, independent of anything else. A quantity is characterized by the laws into which it enters. Do not confuse this with the semantic doctrine attributed to Kuhn and Feyerabend, that the meaning of the word "mass" is fixed by the law-like sentences in which it is used, so that "mass" in a relativistic sentence means something different from "mass" in a Newtonian sentence. Leibniz is talking about mass, not the word "mass." What mass or velocity are, is determined by the laws of nature of mass or velocity. Were nature truly Newtonian, mass would be Newtonian.

The bearing of this on our problem is as follows. *Whether or not a quantity is real or merely phenomenal depends on whether or not the laws in which it occurs are real or phenomenal.*

Quantity and Measurement

Even today we can confuse a physical quantity and its measurement. We have been schooled into thinking that a preparation of whole milk and yoghurt culture has a definite temperature, which may be measured in Fahrenheit, or Celsius, or Kelvin or whatever. "The temperature" is perhaps an equivalence class of such measurements. We distinguish the quantity from its measurement.

To follow Leibniz we should extend this idea to mechanics. Consider the motion of a body, and its *vis viva*. To measure the motion, we need a frame of reference, which will include a system of directions, and also a scale. We may then represent the motion by the vector velocity in a scaled frame of reference. The same frame and scale will enable us to represent the *vis viva*. But the vector v and the energy mv^2 are not the motion and *vis viva*, but representations and measurements of it. We may think of the temperature of the yoghurt mixture as a property of the stuff, which may be variously represented and measured. We should, by extension, think of motion and *vis viva* in the same way. In one frame and scale an object has 0 velocity and energy. In another it has a positive velocity due north and a corresponding energy.

Leibniz's insight is that there is a fundamental difference between motion (represented by a vector v in some system) and *vis viva* (represented by mv^2 in some system). The laws concerning *vis viva* are "real laws," namely the conservation laws, invariant laws. The laws concerning position, its first vector derivative, velocity, and its second vector derivative, acceleration, are merely phenomenal laws, none of which can be regarded as literally true.

Why *Vis Viva* is a Real Quantity of Mechanics

Leibniz has the following conjecture. The law of conservation of *vis viva* is empirically adequate. There is no other equipollent (but incompatible) law of the same phenomena, so the law is invariant. Moreover *vis viva* is not characterized by any other merely equipollent laws of other phenomena.

If this conjecture is correct, the laws of *vis viva* are *prima facie* candidates for real laws of nature, and *vis viva* is a *prima facie* candidate for being a real quantity.

I write in this slightly tentative way of conjectures and *prima facie* candidates because kinetic energy does occur in laws of which Leibniz had no inkling, and consideration of their invariance would take us too far afield. I am here concerned with the meaning of Leibniz's claims about reality, and with their final plausibility, not with their applicability to a later body of physics.

Why Motion is Not a Real Mechanical Quantity

Even Feynman's elementary example makes plain that similar conjectures are false of acceleration. We have see that the law $F = ma$ is not invariant. Hence on Leibniz's criterion, acceleration may be a well-founded phenomenon, but it is not real. Hence velocity and time cannot be jointly real, for the rate of change of a real quantity (v) with respect to another real quantity (t) must itself be real, contrary to what we have estalished about acceleration. Could velocity, however, be real while time was unreal? If so, directed distance (s) would have to be unreal too, for otherwise time, namely s/v would be real. But velocity is only rate of change of directed distance in time, and if both the latter are phenomenal, velocity must be phenomenal too.

Notice that these observations have nothing to do with rotation. I shall soon turn to the bucket experiment, by which Newton sought to prove that rotary motion exists absolutely. Even if Newton were invincible on that score, Leibniz would still have provided a criterion for showing that motion is only phenomenal, not real.

Concealed Metaphysics

Since I have written in terms of elementary mechanics, it may seem as if we have an argument free of metaphysics. That is an illusion. For example it might be protested that *vis viva* occurs in a batch of equipollent laws. After all, $v^2 = 2as$, when s is the distance moved from rest at constant acceleration a. Hence:

$$vis\ viva = mv^2 = 2mas.$$

Feynman's example shows at once that this is not invariant. The Leibnizian retort is that we already know that the measure of *vis viva* is different in different frameworks. $2mas$ is a measure of this quantity, but does not describe the action or interaction of the body whose *vis viva* is being measured. The distance that a body has moved from rest is not an intrinsic property of the body, but only a way of measuring the *vis viva* within a certain coordinate system. The laws of conservation, however, are about the quantities internal to the bodies that interact. With $F = ma$, the force is supposed to be the cause of the motion, but this is not invariant, and hence it is not a real cause but only a phenomenal one. It is just as phenomenal, says Leibniz, as centrifugal (and Coriolis) force.

Thus we see that the power of these reflections, for a mind like that of Leibniz, rests in part on metaphysical notions of cause, and metaphysical distinctions between internal and external properties. In my opinion these notions and distinctions are weighty. The fact that they involve metaphysics does not vitiate my explication of Leibniz. It reminds us only that the explication does not reside solely in the domain of mechanics.

Rotary Motion

I understand Newton to have argued as follows. There can be no proof of absolute translational motion, because for any such motion assigned to a body, we can always choose a frame of reference in which that body is at rest. So we cannot directly disprove the thesis that motion, and hence space and time, are merely relative. However the same trick cannot be applied to rotary motion. There is an intrinsic difference between a suspended bucket that is rotating, and one that is still, and an experiment can determine whether the bucket is still or rotating. This makes sense only if we suppose an absolute space within which the bucket is rotating. Hence absolute space is vindicated.

Before turning to the bucket I should recall Leibniz's short answer to this argument.

> ...it can be understood why I cannot support some of the philosophical opinions on this matter of certain great mathematicians, who admit empty space and seem not to shrink from the theory of attraction but also hold motion to be an absolute thing and claim to prove this from rotation and the centrifugal force arising from it. But since rotation arises only from the composition of rectilinear motions, it follows that if

> the equipollence of hypotheses is saved in rectilinear motions, however they are assumed, it will also be saved in curvilinear motions.[9]

That sounds like a knock-down argument. All rotational motion reduces to linear motion. There is no absolute linear motion, hence no absolute rotational motion. But the argument cuts both ways. If Newton convinces us that there is absolute rotational motion, then consider a point on the tangent of a rotating bucket. This point will have momentary linear velocity at a tangent to the bucket, and if the rotational motion is real, then this linear motion must be real too. Hence we have to be careful in applying Leibniz's "knock-down argument" to the bucket.

The Bucket

Newton suspends a bucket on a long rope. He twists the rope tight. The bucket is held still until the water is level and at rest. Then the bucket is released. At first the bucket spins and the water stays still. But gradually the friction at the perimeter, and viscosity of the water, cause the water to begin to rotate. The centrifugal force of the spinning water drives it to the edge of the bucket, where it piles up, so that the water rises on the perimeter and sinks in the centre. Then Newton grabs the bucket, bringing it to a sudden stop. The water goes on spinning, and only gradually does it slow down, when its level subsides to normal.

Note that following Leibniz I described this in terms of the fictional centrifugal force. A Lilliputian floating in the bucket might describe things in this way: a mysterious force is sucking water to the side of the bucket, but the Lilliputian can opt for the conjecture that the whole bucket is in motion, and the rise in the water is due to the rotation with only centripetal forces (of viscosity) and frictional forces at work. That is the conclusion Newton would urge on him. There is no centrifugal force, only a rotating bucket seen from a stationary position in space.

Leibniz asserts that he can resolve rotary motion into linear motions. We can arbitrarily designate any point on the bucket as at rest. Naturally we cannot designate all as at rest. Here are two pictures, with three vectors of instantaneous motion on the edge of the bucket on a Newtonian scheme, and then as seen when point *A* on the rim is regarded as at rest.

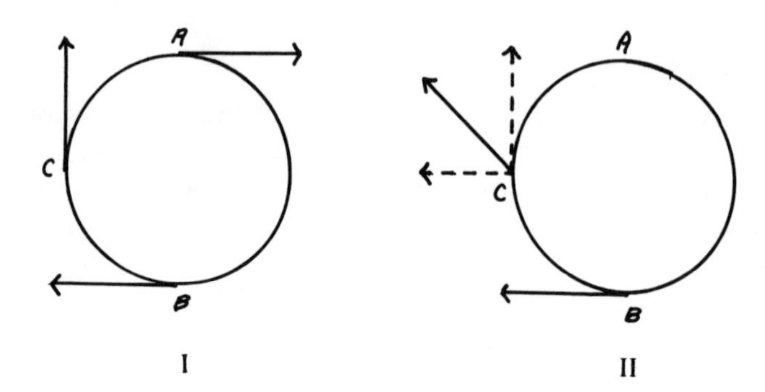

In frame II, the bucket is represented as rotating around the stationary point A on the rim. But then the water would not uniformly sink in the centre and rise on the edges! That response is confused. On our "Newtonian" reference frame, when we rotate a bucket around A, the water rushes towards B, I suppose, and probably slops over the side. but what if we are in the reference frame in which A is stationary? Then the water still rises uniformly around the perimeter, but a person at A must use a modified law of nature in order to describe the phenomena. The situation is exactly analogous to Feynman's Moe and Joe.

Case 1. Newton's frame: we use the simplest laws of motion and compute the rise of water on the edge of the bucket.

Case 2. The frame of a Lilliputian afloat on the bucket: There is a centrifugal force drawing the water to the edge of the bucket. As Ernst Mach remarked, this is not completely unintelligible. The Lilliputian may propose that the fixed stars have begun rotating around his tiny bucket world, and attract the water by further laws of motion and attraction.

Case 3. The reference frame with A at rest: Here is Feynman's Moe with a yet more complicated law of nature required to compute the uniform rise of water to the edge.

It might be objected that Moe's computations, in case 3, will be complex to the point of absurdity. Not so. That is the force of Leibniz's own observation, that a situation like case 3 is only a transformation of case 1, and the computations are no more complex. What is true is that we can see, qualitatively, what curve must be assumed by the water of the spinning bucket after it has reached equilibrium. It will be a parabola (almost). It does not feel as if there could be a similar immediate insight in the case of Moe. On the other hand our "insight" into case 1 results in part from three centuries of training in the Newtonian coordinate system.

Incidentally we should notice that although Newton asserts that he did his bucket experiment, we must presume that he did so only in a qualitative way, and did not actually determine the experimental steady state curve of the water. We have faith that standard Newtonian physics in the usual frame of reference is adequate to the phenomena. So far as I know, this has never been demonstrated experimentally.

Laws of Nature

Most students suppose that the bucket experiment must be central to any consideration of Leibniz on space and time. I believe it is entirely supplementary, and that it does not teach much that is novel. What I propose as new in the present paper -- or rather, as newly revived from Leibniz's time -- is the criterion of invariance which decides (*prima facie*) which quantities are real and which are phenomenal. My observations about the bucket fit into a standard pattern that I have developed elsewhere, in connection with Leibnizian space, and also in connection with the identity of indiscernibles. Leibnizian doctrines go hand in hand with what in recent times we call the underdetermination of theory by phenomena. Even if we consider not only all observed phenomena, but also all phenomena that we could observe, there will be a class of equipollent laws of nature that fit all the phenomena. We can always choose empirically adequate laws of nature that enable us to preserve the identity of indiscernibles.[10] For artificial organizations of monads we can say that the "natural" ordering in space is that associated with the simplest of equipollent laws, which at best assures us that space is only a well-founded ordering of phenomena.[11] Likewise we are not obliged to say that the water in the bucket is in uniform rotary motion. That is merely the simplest description associated with one of innumerable equipollent laws any of which fit the facts.

Gravity

Leibniz certainly had a relational theory of space and time, but it is wrong to say that in any serious way he anticipated the special or the general theory of relativity. It is also a mistake to imagine that he had any influence on Mach's new critique of absolute space and time, or on Einstein's subsequent reflections. However he had a remarkable "nose".

Recall that he lumped gravity with centrifugal force. Recall his contempt for the Newtonians who thought there are real forces of attraction. He clearly though that gravity is a pseudoforce. This is one way to understand the general theory of relativity. I conclude by quoting Feynman once again:

> One very important feature of pseudo forces is that they are always proportional to the masses; the same is true of gravity. The possibility exists, therefore, that *gravity itself is a pseudo force.* Is it not possible that perhaps gravitation is due simply to the fact that we do not have the right coordinate system? After all, we can always get a force proportional to the mass if we imagine that a body is accelerating. For instance, a man shut up in a box that is standing still on the earth finds himself held to the floor of the box with a certain force that is proportional to his mass. But if there were not earth at all and the box were standing still, the man inside would float in space. On the other hand, if there were no earth at all and something were *pulling* the box along with an acceleration *g*, then the man in the box, analyzing physics, would find a pseudo force which would pull him to the floor, just as gravity does.
>
> Einstein put forward the famous hypothesis that accelerations give an imitation of gravitation, that the forces of acceleration (the pseudo forces) *cannot be distinguished* from those of gravity; it is not possible to tell how much of a given force is gravity and how much is pseudo force.
>
> It might seem all right to consider gravity to be a pseudo force, to say that we are all held down because we are accelerating upward, but how about the people in Madagascar, on the other side of the earth -- are they accelerating too? Einstein found that gravity could be considered a pseudo force only at one point at a time, and was led by his considerations to suggest that the *geometry of the world* is more complicated than ordinary Euclidean geometry.[12]

But of course Einstein's ideas do not stop at gravity. Here is one philosopher's passing comment on where we go:

> Einstein wanted to claim genuine reality for the central theoretical entities of the general theory, the four-dimensional space/time manifold and associated tensor fields. This is a

serious business, for if we grant his claim then not only do space and time cease to be real, so do virtually all of the dynamical quantities. Thus motion, as we understand it, itself ceases to be real.[13]

Department of Philosophy
and
Institute for History and Philosophy of Science
University of Toronto

Notes

Abbreviations:

GM for Gerhardt (ed.) *Leibniz: Mathematische Schriften* (Berlin: 1875-1890, seven vols.)

Loemker for Loemker (ed.) *Leibniz: Philosophical Papers and Letters*, (Dordrecht: Reidel, 1969)

1. R.B. Lindsay, *Physical Mechanics*, 3rd Edn., Princeton, 1961, p. 381.
2. *Specimen Dynamicum* (1765), GM VI, p. 238, Loemker, p. 438.
3. *Ibid.*
4. For popular expositions of this effect, described in most middle level textbooks of mechanics, see M. Correll, "The Case of the Coriolis Force," *The Physics Teacher*, Jan. 1976 or J.E. McDonald, "The Coriolis Effect," *Scientific American*, May 1952.
5. D. Halliday and R. Resnick, *Physics, Parts I and II, Combined*, 3rd. Edn., New York, 1976, p. A3.
6. R. Feynman, *Lectures on Physics*, Berkeley, 1962, Vol. I, p. 12-11.
7. G.M. II, p. 193; Loemker, p. 419.
8. For the "small paper sent to Mr. Viviani, and which seems appropriate to persuade the gentlemen in Rome to allow the Copernican hypothesis" (cf. previous footnote reference), see G.M. VI, 144-7.
9. *Specimen Dynamicum*, G.M. VI, p. 253; Loemker, p. 439.

10. Ian Hacking, "The Identity of Indiscernibles," *The Journal of Philosophy*, 1975.
11. Ian Hacking, "A Leibnizian Space," *Dialogue*, 1975.
12. Feynman, *ibid.*, p. 12-11f.
13. Arthur Fine, "The Natural Ontological Attitude," forthcoming in J. Leplin, ed., *Scientific Realism*, Notre Dame.

Graeme Hunter

MONADIC RELATIONS

F.H. Bradley may well have been the last major philosopher to acknowledge that relations, in their own right, pose an authentic metaphysical problem. Near the beginning of *Appearance and Reality* (p. 21) he gives this problem a memorably pessimistic formulation:

> Relation presupposes quality, and quality relation. Each can be something neither together with, nor apart from, the other; and the vicious circle in which they turn is not the truth about reality.

In this paper I shall be concerned only with the dependence of relations[1] on qualities and not the converse dependence, though each is recognized by Leibniz. On the matter of concern, however, Bradley's formulation is again as good a point of departure as any (p. 27/8):

> But how the relation can stand to the qualities is, on the other side, unintelligible. If it is nothing to the qualities, then they are not related at all; ...But if it is to be something to them, then clearly we now shall require a *new* connecting relation. ...And, being something itself, if it does not itself bear a relation to the terms, in what intelligible way will it succeed in being anything to them? But here again we are hurried off into the eddy of a hopeless process, since we are forced to go on finding new relations without end.

Let's take a concrete example. Perhaps I am eight feet from that chair. Now suppose I require a metaphysical explanation of that relational fact. It seems that there are two possible kinds of explanation. One kind involves only two terms -- me and the chair. But it does not seem

K. Okruhlik and J. R. Brown (eds.), The Natural Philosophy of Leibniz, 151–170.

plausible that even the most exhaustive analysis of my qualities (i.e. my non-relational properties) or of the qualities of the chair, or of both, would ever provide the metaphysical explanation of my being eight feet from the chair. Why not? Well, intuitively because we think that a scrupulous analysis of my qualities or those of the chair would produce identical results, whether I was eight feet, eight inches or eight miles away from it and that therefore no such analysis could provide any explanation of why things are as they happen to be as regards my spatial relationship to the chair. So much then for the two-termed analysis.

The alternative explanation would involve three things -- me, the chair and the relation of 'being eight feet from', where the latter expression denotes something beyond any qualities of mine or the chair's. It is the connection of the two terms (me and the chair) with the relation, 'being eight feet from', which is to explain this state of affairs. But now the question arises 'Just how are these three things connected?' If they are genuinely three things, then some explanation of their *connection* is called for. So the three-termed metaphysical explanation turns out to be as much of a dud as the two-termed one: In order to explain how it can be that I am eight feet from the chair, *it* says that I am connected to the chair by the relation of 'being eight feet from'. But of course this is simply to explain the dyadic relation 'x is eight feet from y' by the *triadic* relation 'x is connected by R to y'. And since what we want is an explanation of a relational state of affairs, to explain it by a further and more complex relation is circular. This is reminiscent of the famous 'third man' argument with which Parmenides puzzles the young Socrates.

It is this apparent failure of both the two-termed and the three-termed explanations to do justice to the question of dyadic relations, together with the appearance that no better explanation is possible, that I shall call the *metaphysical* problem of relations, I shall contend that what Leibniz has to say about relations should be regarded, first and foremost, as an answer to this problem.

But in addition to this central, metaphysical problem, that must be faced by any theory which takes metaphysics seriously, Leibniz's remarks on relations are shaped by two further difficulties which I shall characterize rather loosely as 'historical'. The first is analogous to the Neoplatonic problem of the relation of God to creation, and probably arises out of Leibniz's peculair adaptation of Neoplatonism. The second was thrust upon Leibniz by his time. Descartes had argued powerfully for the real distinction of mind and body, but had only superficial things to say about their interaction. For Leibniz, what demanded explanation was not just the interaction of mental and physical substances but, more broadly, the interaction of substances in general.

Most commentators, following Russell, have tried to construe Leibniz's occasional remarks on relations as by-products of his logical/grammatical view that relational predicates (or oblique cases) could be reduced to qualitative predicates (or nominative cases). And there are some obvious ways in which his logical initiatives and his metaphysical views were connected, so that much of the work done on Leibniz's theory of relations by people under Russell's influence -- and that is almost everyone in the Anglo-American sphere -- is indisputably useful in illuminating these connections. However, associating his logical and metaphysical pronouncements in this way has the drawback of making the metaphysical problem seem as marginal as its logical counterpart, or even, on certain views of logic, something of a wild goose chase. (See e.g. B. Russell, p. 226.) But a survey of the philosophical writings will reveal that Leibniz discusses relations in many contexts other than grammar or logic, e.g. respecting ontology, the phenomenal/real distinction, categories of being, perfection, solipsism, space and time, perception, action and passion, and others. In all of these the primary motivation is metaphysics, not logic, and the underpinnings of the account are those which at the same time provide Leibniz's answer to the three fundamental questions which I outlined above.

I. The Metaphysical Question

Some relations don't present any metaphysical problems or at least none of the ontological kind with which I shall be concerned here. For example, if you are taller than I am, then there is a relation between us. But it does not seem to be a metaphysical swindle to say that the explanation of this relation is simply that my height is 6'2", yours is 6'7", and that the 'relation' between us is something which the mind creates when it compares our heights.[2] The relation, however complex its logical structure may be, is simply a 'being of reason' which arises from the act of comparing my height and yours, and need be accorded no place in the ontology.

This sounds suspicious on first hearing, since our simplest (and perhaps soundest) semantical intuitions lead us to expect that any true statement S be a straightforward verbal representation of the fact in virtue of which

S is true. Hence, if S is a true relational statement, the fact to which it corresponds must be itself relational. However Leibniz's semantical reflections lead him to a somewhat more complicated semantics. The relational statement 'aRb' is triply ambiguous. It may be conceived to be 'about' a or b or both, i.e. either a or b or both can be its subject. If it is thought to be about a, then it is to the qualities of a that we must look for its truth-conditions, the same, *mutatis mutandis*, where b is subject. If both are considered as subject, however, then the relation arises as an independent entity connecting them. But this relation "being neither substance nor accident, must be a purely ideal thing" (GP VII, 401 = L, 704). In general for Leibniz relations require and have a foundation in the category of quality ("indigent fundamento sumpto ex praedicamento qualitatis" C,9 = PP, 133/4) and to any given ontological foundation corresponds a unique relational superstructure. For example, given our respective heights, no other appearance is possible than that of your being taller than I am. But the relation is nothing more than an appearance for all that. For Leibniz, our having different heights does not really *connect* us. Thus, among other names, Leibniz calls such relations 'relations of comparison' (see e.g. NE II, xi, 4) which he distinguishes from those of connection (C, 434/5). Other such relations are those of *similarity* in some point, e.g. if you and I are of the same height, or *identity*, as e.g. Phosphorus' being Hesperus. John Earman, who has written one of the finest articles on this topic, calls such relations "second-order", because they depend upon non-relational or first-order *qualities*. These first-order qualities, departing now from Earman's terminology, characterize individuals in a way that is independent of the mind's activity. To stick with the old example, if you are taller than I am, the relation 'is taller than' is partially a construction of the mind, but it is also partially dependent on my "first-order" property of being 6'2" and yours of being 6'7". But those qualitative properties, in this context at least, are taken as data independent of the mind's activity. Such "second-order" relations then are not the problem. They are relations which, while handy to have around, are not, and do not have to be, part of our ontology.

The difficult kinds of relations are not those which introduce a comparison based on similarity or difference, but those which introduce a genuine *connection*. These may be causal, spatial, temporal or intentional, and perhaps there are other categories as well. If I wrestle you, bump into you, shake hands with you, if I am a foot, a yard or a mile away from you, if I am you successor or predecessor at this chair, if I see you, hear you, or get to know you, in each case there seems to arise a relation which cannot be ascribed to any combination of first-order qualities, together with the activity of the mind.

The problem in relations like seeing, hearing or knowing is familiar to all, because it is much discussed in another context, that of *intentionality*. For me to see you, it is insufficient that you be in front of me, for me to have my eyes open and to have a mental image of you. All that could be true if I were blind and my mental image of you were produced by the hidden gadget of my controlling Martian -- all these prerequisites of my seeing you could obtain, and yet I could not truly be described as seeing you. Part of the requirement seems to be that your presence *cause* me to have the image I have, in other words there must be a real, first-order relation between you and me in order for any genuine case of my seeing you to take place. Similarly, in a non-intentional context, if I bump you or stand near you, there can be no relation-free description of that state of affairs built up uniquely out of first-order qualities, no matter how ingeniously the selection of them is made. Nor can the situation's relational character be explained by appeal to the mind's activity. For if I bump you, this is very different from the mind's conceiving that I bump you, together with the simultaneous occurrence of what we ordinarily take to be our respective first-order properties. And this difference must be represented in any adequate theory. Thus the real problem is first-order, not second-order relations or, to use a fitting pair of Leibnizian terms, it is relations of causality, not those of similitude. (Leibniz uses these terms at Ak VI, vi, 456 = GP IV, 175.)

But first-order relations must be especially difficult for Leibniz. We have only to recall that monads 'have no windows', i.e. are without causal connection to any other created monad (Monadology, sect. 7)[3] to see that Leibniz should encounter even more difficulty than usual here. You and I are created monads, so the problem for Leibniz can be stated simply as a variant of the earlier example: How is it possible for me to be eight feet from you?

To say that you and I are monads is to say, in the last analysis, that you and I are spiritual, atomic substances (see GP IV, 48s/3 = L, 537), which are defined, as he tells de Volder, by perception and appetite (GP II, 270 = L, 537). Our perception, of course, cannot be strictly causal in nature, since, if it were, the monads would not be windowless. Instead it is *representational* in nature. That is to say, the monad's perception is its ever-changing *activity* of world-representation (G IV, 572 = L, 579). (For a fine discussion of this, see McRae I, 63ff.) Such representation is always perspectival, from the 'point of view' of its phenomenal body (Monadology, sect. 62), with the result that each monad is like a separate world, or better, a separate world-view, in a totality which consists only of world-views. The unity or connection among the monads of a world is a function of the similarity, in form and content, of their respective world-respresentations. Indeed, for most purposes, each Leibnizian monad may be understood as an ordered pair, whose first member is a perspective and whose second member is a set of objects. A Leibnizian world is the set of monads which agree on the objects represented, while disagreeing on the perspective from which they are represented. To see both how this works and how it connects to our problem, let us return to the case of my being eight feet from you: Our monads are unconnected. Nevertheless, each of us represents the *same* world, save for differences in perspective or, put more carefully, no difference can be detected between your representation and mine, which is not a difference in point of view. Both of us have representations of ourselves and the other in our world-context, both agree as to who we are and what distance separates us. Phenomenally speaking, this situation consists of only you and me separated by a certain distance and this is not provided with any metaphysical analysis. For our apparent relationship, to echo Bradley, 'is not the truth about reality'. Leibniz presents no metaphysical analysis of phenomenal relations because they would not support it. Phenomenal relations are, of course, the subject of some logical investigations, but that is not at issue here. However intricate the logic of relations may be and however inadequate Leibniz's analysis of it, relations are only apparent existents and cast no shadow across the ontological infrastructure from which they arise. They result from the confused perception of reality with which the senses systematically deceive us.

At the noumenal level, there is only a consensus among the monads, in particular between you and me, a congruence or fitting of our representations. Thus, at the noumenal or monadic level, the connection between us is explained by similarity, one of the second-order relations, which does not appear in the ontology. The second-order similarity in what is represented is based upon the first-order quality of our representing as we in fact represent. The net result is this: At the level of phenomena, relations are a primitive, irreducible fact. At the level of monads, i.e. in reality, there are no relations at all. This is why I go along with Prof. Ishiguro and others who have contended that the several attempts to argue for a Leibnizian 'reduction' of relations, have been in vain. The world of related bodies is preserved in monadic representation. But monadic representation is itself qualitative, not relational.

II. The Neoplatonic Problem

In offering the foregoing account I have probably let myself in for a criticism. I'm sure some people will think that the attempt to analyze first-order relations away by appeal to the notion of 'representation' is doomed to failure from the outset because the predicate 'represents' itself denotes a first-order relation of the intentional variety. Moreover in the case of monadic representations, the proposed model portrays each monad M as representing a state of affairs S, where S itself is an infinitely complex relational state of affairs. Thus it is by the doubly relational formula 'M reprsents S' that I am trying to account for first-order relations. And this difficulty is linked with and leads naturally to a deeper one -- the first of what I earlier called the 'historical' problems of Leibniz's doctrine of relations. It is a problem which Leibniz has in common with the Neoplatonists and was once very neatly characterized in the context by E.R. Dodds (Dodds, p. 16):

> The main difficulty for Neoplatonism, as for any philosophy which affirms a transcendent Absolute, is to prevent the Absolute and the world from falling asunder so that either the former becomes relative or the latter becomes unreal.

Now Leibniz's God is a transcendent absolute (see e.g. GP VII, 303 = L, 487), so Leibniz definitely has that problem. But for him it is also infinitely multiplied, because each monad is related to its own perceptions in a way analogous to the Neoplatonistic God's relation to the world. Thus in one fragment (Grua, 553) Leibniz calls the monad a 'diminutive God-likeness, a universe-likeness eminently'. Again, in the *Monadology* (sect. 60) and in the *Theodicy* (sect. 403), Leibniz reminds us that a monad which was in no way confused in its representations would be a god. The point of similarity is, of course, the following: Just as God gives rise to the monads "by continual Fulgarations from moment to moment" (*Monadology*, sect. 46), so the monads give rise to their world-perceptions by continual fulgarations of expression (See also G IV, 553). Now the problem, to rephrase Dodds, is this: If the monad is a true unit, transcending the multitude of its perceptions, as is implied when Leibniz speaks of it as being 'eminent' with respect to the represented universe (Grua, 553, Theod, 403), then the genuine multiplicity of its universe seems threatened. Conversely, if this multiplicity is enshrined as a principle from the outset, it is difficult to see how the monad can remain innocent of relations of connection.

I shall emphasize three facets of this problem of intra-monadic relations. First there is the question of how to reconcile the unity of the monad with the diversity of a given perception. Next there is the problem of the unity of the monad given its *changing* perceptions over time. Lastly whatever arguments are used to show the unity of the monad must be checked to see if they also permit the diversity of its perceptions. Simply put then, Leibniz seems to be faced with Hobson's choice: Either admit real intra-monadic relations after all or deny the reality of the monad's perceptions. Either option means the failure of Monadism.

I know of no single text in which Leibniz addresses himself to this problem or provides a definitive answer to it. However, scattered among his many remarks about relations are, as it seems to me, some clues as to how such an answer might have gone. I would like to attempt a brief reconstruction of it here.

The problem is to find within Leinbniz's characterization of the monad some concept or concepts which allow for real diversity and real change among monadic perceptions, while preserving the transcendent unity of the monad itself. Neither appetite nor representation, the two characteristic aspects of the monad, provide any such concept immediately, but the latter rewards closer inspection. The term 'representation', like many words ending in 'tion', is ambiguous, meaning sometimes an activity and at other times the result of that activity. Leibniz uses the term in both these senses. When he speaks of representation *functionally*, a usage very well documented by Kulstad (p. 58), he generally intends the result rather than the activity of representation. And there is a host of other passages, equally well documented by Dillman (p. 313), in which the emphasis is reversed. It is by placing weight -- I hope not too much weight -- on the term's active sense, that I will try to reconstruct the Leibnizian answer to the question of unity and diversity.

There is a kind of represetnation which is an activity, with which we are all familiar, and which it is sometimes useful to keep in mind as a model, though in many ways imperfect, for monadic representation. In

Leibniz's time it was common in both English and French to speak of the 'représentation', meaning 'performance', of a play, and it has remained a common sense of the French 'representation'. Taking this sense as a model is surprisingly fruitful in understanding how Leibniz imagined that the many could be represented in the one, without, for all that, the one being multiplied.

To begin with, take a moment, a time-slice, of a play we know: Perhaps the stage is empty except for Hamlet and Claudius -- Claudius is praying, Hamlet debating with himself whether to kill Claudius now. What is it which unifies the dyad, Hamlet/Claudius, which causes us to see them as being elements of a single event? Surely it is this: That they together constitute a moment in a larger whole -- the *Hamlet*-representation. It is only in virtue of this representation that we conceive of them as connected at all. And for us, God-like, penetrating observers that the theatre allows us to be, this time-slice of the representation contains traces of all the effects of what went before and of all the events of which it is about to be a cause. Thus the unity of the momentary scene is dependent on the prior unity of the ongoing representational activity, from which the time-slice is an abstraction. In general, the unity of an activity is prior to the diversity of its moments.

Is that convincing? Perhaps not entirely. But at least the foregoing sentences seem to recast in the familiar and more intuitive model of the play what Leibniz suggests darkly to Bayle, when he writes defending the notion that the soul, indivisible as it is, can envelope a multitude of traces at any moment (GP IV, 551/2):

> And since there is a huge amount of diversity in the present state of the soul which *knows* a great deal and *feels* infinitely more, and that these present diversities are an effect of those of a preceding state and a cause of those of a future state, we thought we could call them *traces* of the past and future, in which a spirit of sufficient penetration could recognize the one and the other...

Now even granted that momentary representational time-slices are absractions and that such unity as they have is a function of the on-going activity of representation, there remains the question of how this activity can be a unit, given that it has temporal parts. To this Leibniz would answer that a representation does *not* have temporal parts and that, if it did, it would not be a unit. As Leibniz tells both de Volder and Rémond, souls or monads are the *foundations* of time; they are not in time (GP II, 253 = L, 531; GP II, 268f = L, 536f; GP III, 623).

How can this be illustrated on the more familiar model of the play? Someone might point out that it is divided into scenes and acts and bound by a certain performance time. But this is the play, regarded as an artifact of *our* world. The divisions and performance time are related to *our* time. Yet as spectators we become familiar with another world and time, a time in which we do not live ourselves, yet in which the characters of the *Hamlet*-story appear to live. It is a time analogous to, but not identical with, our own. It is analogous to ours in that, e.g., its characters grow older as time goes by, what is past cannot be changed or relived, its future branches with alternatives, and so on. But it is not our own time. In our time, the *Hamlet*-world is only exposed to us for several hours; in the time of that world it is several months. And if we look closely at the time of the *Hamlet*-world, we see something to be true of it which is exactly analogous to Leibniz's view of the relation of monads to time. In the *Hamlet*-world, the characters all *appear* to be immersed within the time of their world. But as spectators we know, though we may thrust the knowledge aside while absorbed in the play, that the illusion of time, which the play creates, depends entirely on the representational activity of the actors and would vanish but for that activity. Time, in the *Hamlet*-world, is an abstraction from the activity of the denizens of that world, just as, in the real world, according to Leibniz, time is an abstraction from the changing sequences of monadic representations. (The idea that time, for Leibniz, is dependent on change is developed by R. McRae in McRae II.) In the play analogy, the representation of the actors takes place *within* the time of our world, and is thereby able to be the foundation of the time of the *Hamlet*-world. In our world, the representation of the monads takes place in eternity and thereby serves to found our time. Without this monadic foundation, time, phenomena and the story of our world would vanish as quickly and tracelessly as the well-ordered dream to which Leibniz sometimes likes to compare them. But if there is not time in the noumenal world, then there can be no earlier and later, no relations of earlier and later, and therefore no intra-monadic temporal relations.

Thus the claim that there are no relations in the monadic realm which is the ontic infrastructure of our world, reposes on the analysis of *representation* as *activity*, and of activity, in turn as a principle of unity outside of time, and time's real foundation. This I take to be the conceptual skeleton of Leibniz's answer to the question of how the represented many cohere in an unified *representans.*[4]

But even conceptual impeccability, now generously assuming the above account of representation to enjoy it, would not be a sufficient guarantor of coherence. One has only to recall Leibniz's criticism of the ontological argument to see this. Leibniz would be the first to insist on some proof that the notion of a simple substance capable of envoloping a multitude within its own simplicity is coherent.

I have been unable to find any attempt of an *a priori* proof of the possibility of this notion. Leibniz never seems moved to produce an *a priori* argument, given that, in his eyes, a knockdown *a posteriori* argument was available, as well as a most enlightening analogy.

The *a posteriori* argument to which I referred is found in *Monadology*, sect. 16, where Leibniz indignantly criticizes Bayle for having found fault with the notion that a simple substance might have multiple affections. Leibniz writes (*loc. cit.*):

> We ourselves experience a multitude in a simple substance when we find that the most inconsequential thought of which we are conscious enfolds a variety which is in the object. Thus all those who recognize that the soul is a simple substance must recognize this multitude in the Monad.

He had put it still more plainly a quarter of a century earlier to Arnauld (GP II, 112 = L, 339):

> The possibility of a just representation of many things in a single one cannot be doubted, since our soul provides us with an example.

The *a posteriori* argument then is one which depends on the immediate access which we have to our own thoughts, in which simultaneous simplicity and multiplicity can be discovered.

In the "Principles of Nature and Grace", sect. 2, he argues for the same conclusion by analogy with the way in which a focal point may be the centre of an infinity of lines. The focus expresses many relations, e.g. the relation of being coterminous for infinite pairs of lines, while itself remaining simple. Incidentally these two ideas are combined in a very elegant German verse which Leibniz wrote illustrating his monadic philosophy. (Cited in *Mahnke*, p. 17.)

These two ideas, particularly the *a posteriori* argument, provide the missing possibility proof which, in conjunction with the analysis of representational activity, allows Leibniz a formally, if not intuitively, satisfying proof that there are no intra-monadic relations at the monadic level.

But even this is not quite the end of the Neoplatonic problem. For recall that Dodds' sword had a double edge -- the trick is not simply to show how what is many is, in some important sense, one; it is rather to accomplish that while still preserving the genuine multiplicity of the many.

The key to a representational multiplicity which does not threaten monadic unity is the Leibnizian notion of the 'well-foundedness' of an appearance. I shall sketch this briefly and then show how it applies.

The well-founded is a middle ground between the real and the merely apparent. To say that something is well-founded is to say that it has a basis in the monadic infrastructure. (Comp. GP III, 623.) Well-founding can be characterized almost recursively as follows: Animate bodies are well-founded because at the monadic level, there is some monad which represents the world from the point of view which that body would have, were the phenomenal world real. (Comp. Monad, sects. 62, 63; PNG, sect. 4.) Inanimate bodies are aggregates of animate ones, each of which is well-founded in the way just mentioned (G II, 439). Phenomenal events of this world then, are well-founded provided all the monads of this world represent them as occurring.

Now this allows Leibniz to account for the multiplicity within the representations of a monad in such a way that this multiplicity is neither full-blooded enough to compromise monadic unity nor so diluted as to compromise genuine diversity.

To take an example, part of the reason why my current representation of being in front of a number of people is accurate is that, corresponding to each person of whom I am aware, there is a monad representing to itself the same event from the perspective which, according to my representation, that person ought to have. Yet the fact that these *other* monads are doing this, while conferring a kind of objectivity upon my pluralistic perception, does not threaten its unicity. There is a genuine multiplicity which provides an objective foundation for my perception, but since it lacks causal connection with my perception, it poses no threat to its unity. In this way the problem of intra-monadic relations, which I have styled the 'Neoplatonic problem; has a coherent, if rather strained, solution in Leibnizian metaphysics.

Before leaving this problem altogether, the foregoing remarks might be rounded off with a brief comment on relations between God and created monads. After all, it is here that the Neoplatonistic comparison came in: But unfortunately Leibniz is still less forthcoming on this topic than he is on the others which I have been considering and I do not feel in a position to do more than guess at his ultimate view. There are, as I see it, three ways which Leibniz could go and I shall content myself with sketching these.

First, Leibniz could conceive of God by analogy with monads and the world by analogy with monadic perceptions. This is certainly suggested when the universe is said to result from divine "fulgurations" (Monadology, sect. 47), "emanations" (GP IV, 553; Grua, 396), or "outflowings" (Theod., pref. GP VI, 27). But Leibniz (to my knowledge) never develops this doctrine, and this may be becuase he feared having it confused with another view which he despised and took every opportunity to revile, namely that God is the world-soul. (See GP II, 304/5; Grua, 396; GP VI, 529-39 = L, 554-60).

Another line of argument which Leibniz could have taken would be to stress the *difference* of God from creation and the uniqueness of the divine case. In this way, which holds small promise of a satisfying argument, Leibniz might have sought to attenuate, if not dissolve, certain apparent relations between God and creation. Yet despite texts like "On the Ultimate Origin of Things", where Leibniz is unequivocal about God's transcendence (GP VII, 303 = L, 487), to his credit Leibniz does not seem inclined to use this transcendence to cut the Gordian knot with arguments which depend on human ignorance, finitude or other bottom-of-the-barrel inadequacies.

My impression is that Leibniz would favour here, as usual, a middle course in which God's transcendence of and immanence in the created world would come to expression equally, but I have by no means either the textual or the theoretical basis for working out the details. So much then, for the Neoplatonic question.

III. The Cartesian Problem: Inter-Monadic Relations

For Descartes the crucial metaphysical puzzle, for which he could find no satisfying answer, was that of the interaction of mind and body. As with the foregoing Neoplatonic question, this Cartesian one becomes simply a special case of a much more general problem when it is translated into Leibnizian terms. The general problem is that of the communication of substances. That this should be the Leibnizian counterpart of the Cartesian question is quite obvious when one considers that, for Leibniz, bodies are simply aggregates of monadic substances, not substances in their own right (GP II, 135) as they were for Descartes. Thus, when Leibniz publishes his "New System" its full title is: "A New System of the Nature and the Communication of Substances, as Well as the Union between Body and Soul". And in the text of that paper he treats the two aspects of interaction as more or less interchangeable illustrations of the central problem of his own theory (GP IV, 453 = L, 457).

There are some additional problems peculiar to Leibniz's own version of things, which arise from his defining bodies as he did and which spill over into the mind-body question. E.g. there is the question: 'Since the monad's only link to the universe is through expression of it, how is the monad's expression of its body to be distinguished from its expression of the rest of the universe?' I shall ignore this and related questions here. The general difficulty which I wish to consider is that of accounting for causal relations among substances.

Given that bodies are phenomenal and that the phenomenal is well-founded in the sense outlined above, it follows that the phenomenal interaction of bodies is to be explained in terms of the noumenal interaction of monads. And since all is composed of monads, in answering the general question of inter-monadic causal relations, we shall also be answering indirectly the question of mind/body interaction and that of intentional relations, which were part of the original programme of this paper.

For the sake of a simple, intuitive example, suppose I push you. To what features of that phenomenal event must a noumenal account do justice in order to count as an explanation? As it seems to me, they are at least the following: The phenomenal event takes place in time, involves two extended bodies, one of which is active, the other passive, one of which pushes, the other of which offers resistance to the push, then is moved by it.

I shall not try now to develop Leibniz's noumenal account of all these features. Instead I shall concentrate on the one feature which Leibniz himself emphasizes in his metaphysical writings and of which he seems satisfied with his account -- that of action and passion. Common sense tells us that if I push you we enter into a causal relationship as active and passive term respectively. In the monadic metaphysics, this is somewhat problematic since a) monads have no real causal relations and b) monads are entirely active. However, two short sections of the *Monadology*, together with some other considerations, put us on the best way toward an answer to these difficulties. In section 49 Leibniz writes:

> The creature is said to *act* on external things insofar as it has some perfection and to be passive with respect to another creature insofar as it is imperfect.

Consider first that a creature is said to *act* insofar as it is perfect. Consider next that the *imperfection* of a creature arises solely "from the original limitation of the created thing" (PNG, sect. 9 (GP VI, 603 = PP, 200); Monad, sect. 47). These texts together entail that, insofar as the monad is, it is perfect -- a doctrine of no striking originality. But then of course it follows that since the monad's perfection is activity and its perfection is its being, that its being is also its activity. The horizon of its being or perfection simply coincides with the limits of its activity. Now the nature of the monad's activity, as has already been seen, is representation. Its limitation is in the distinctness of its representation. Every created monad represents the entire universe, so there is no *quantitative* limit to its activity. But in large measure this representation is confused (GP VI, 604 = PP, 201), thus introducing a *qualitative* limitation. Only God represents the entire universe distinctly. Perhaps the limitation of monadic expression may be understood by analogy with the limitation of a genre of art. A landscape painting, no matter how faultless and inspired its evocation of its world, is essentially inadequate to certain aspects of its world -- for example the audible and tactile qualities. So the monad expresses its entire world with qualitative rather than spatial horizons.

Now how do these considerations help to explain the phenomenon of my pushing you? Leibniz answers in sect. 52 of the *Monadology*:

> ...when God compares two simple substances he finds in each reasons which oblige him to adapt the other to it, and consequently what is active in certain respects is passive from another point of view: *active* insofar as what is known distinctly in it provides a reason for what is going on in something else; *passive* insofar as the reason for what is going on within it is found in what is known distinctly in something else.

Thus the key to phenomenal action and passion in the creature lies not in the notion of expression itself but rather in that of perfection or *distinctness* of expression. Put in other words, the phenomenal activity and passivity define the qualitative horizon of the noumenal activity of representation. Thus when I push you, the reason for this event is found more distinctly expressed in me than it is in you, so preserving my agency and your passivity at the noumenal level. To the resistance which your body offers, a result of *your* activity, will of course correspond a distinctness in *your* representation, counterpart to a confusion in mine. The continuous changes in our relative motions (during which, of course, the totality of force is preserved) are mirrored in the noumenal realm by changes in the relative distinctness or confusion of our representation, while the overall degree of distinctness at the monadic level remains constant. Thus the action and passion of the phenomenal level have, at the noumenal level, an *explanans* which is compatible with both the mutual causal independence of monads and their, quantitatively speaking, entirely active nature. It is by the good offices of the pre-established harmony that the monads influence one another in a manner which is wholly ideal (*Monadology*, sect. 51).

L.J. Russell was very critical of Leibniz's monadic solution to dynamical questions, particularly as it appears in the latter's correspondence with de Volder. But the main focus of his objection, like that of some others among Leibniz's interpreters, is that the phenomenal, dynamical properties of things are not ultimately real for Leibniz. But here Russell simply differs from Leibniz on a basic point. For Russell, to save the phenomena is to preserve them, for Leibniz it is *rationem reddens*, giving a reason for them.

In summary then, I have tried to outline some general and particular considerations which enabled Leibniz to hold the monadic infrastructure to be entirely free of first-order, causal relations, except to God, while at the same time preserving the plurality and relativity of created beings. As the title of this paper suggests, Leibniz defends a doctrine of 'monadic relations', the paradoxical appearance of which I have tried to remove. To do so has required discussion of monadic activity, representation, well-founding and distinctness. But this is only half the story. The other half involves the sense in which qualities of monads are themselves relational -- a paradox better left for another time.[5]

Department of Philosophy
University of Toronto

Bibliography

Primary Sources

Note: Well-known works, available in many editions and translations are cited only by title. All entries are according to the short form used to identify texts in the body of the paper.

Ak — G.W. Leibniz: *Sämtliche Schriften und Briefe* Berlin: Akademieverlag. Began publishing 1923 and still in progress. Various volumes variously dated.

C — G.W. Leibniz: *Opuscules et fragments inedits* Ed. L. Couturat, Hildescheim: Olms, 1966. Reprint of Paris, 1903.

GP — G.W. Leibniz: *Die Philosophischen Schriften* 7 vols. Ed. C.I. Gerhardt. Hildesheim: Olms, 1978. Reprint of Berlin, 1875-90.

Grua — G.W. Leibniz: *Textes Inédits* ed. G. Grua. Paris: Presses Universitaires de France, 1948.

L — G.W. Leibniz: *Philosophical Papers and Letters* ed. and trans. L. Loemker (2nd ed.). Dordrecht: Reidel, 1976.

NE — *Nouveaux essais sur l'entendement humain* (New Essays on the Human Understanding)

PNG — "Principes de la nature et de la grace fondés en raison" ("Principles of Nature and Grace, Founded in Reason")

PP — G.W. Leinbiz: *Philosophical Writings* ed. G.H.R. Parkinson. London: Dent, 1973.

Theod — *La Théodicée* (*The Theodicy*)

Secondary Sources

Bradley, F.H., *Appearance and Reality* Oxford: The Clarendon

Press, 1968. Reprint of Oxford, 1893.

Dillman, E., *Eine neue Darstellung der Leibnizischen Monadenlehre* Leipzig: Reisland, 1888.

Dodds, E.R., *Select Passages Illustrating Neoplatonism* Chicago: Ares, 1979. Reprint of London, 1923.

Earman, J., "Perceptions and Relations in the Monadology" *Studia Leibnitiana* 9 (1977) pp. 212-230.

Ishiguro, Hidé, "Pre-established Harmony *versus* Constant Conjunction" *Proceedings of the British Academy* vol. LXIII (1977) pp. 239-263.

Kulstad, M., "Leibniz's Conception of Expression" in *Studia Leibnitiana* 9 (1977) pp. 55-76.

Mahnke, D., *Unendliche Sphäre und Allmittelpunkt* Stuttgart-Bad Cann-Stadt: Frommann, 1966. Reprint of Halle, 1937.

McRae, R., *Leibniz: Perception, Apperception and Thought* Toronto: University of Toronto Press, 1976. II: R. McRae: "Time and the Monad" *Nature and System* I (1979) pp. 103-109.

Russell, B., *Principles of Mathematics* New York: Norton.

Russell, L.J., "The Correspondence between Leibniz and de Volder: in *Leibniz: Metaphysics and Philosophy of Science* ed. R.S. Woolhouse. Oxford: Oxford University Press, 1981. pp. 104-118.

Notes

1. I have taken dyadic relations as paradigmatic throughout. Examples involving polyadic relations for my purposes would have increased the paper's opacity with no corresponding increase in depth. No attempt has been made to disguise the

fact that the paper was originally written for oral presentation, though some additional and more technical comments have been added here and there.

2. I am perfectly prepared to admit, at least for present purposes, that this analysis of the asymmetrical relation 'is taller than' *would be* a swindle, if proposed as an analysis of the *logic* of the relation. Russell (*Principles of Mathematics* pp 221-224) showed quite some time ago the futility of analyzing any asymmetrical relation "xRy" into the non-identical properties "P(x)" and "P'(y)", where "P" and "P'" are themselves free of asymmetrical relations. The reason is that the *analysans*

$$P(x) \ \& \ P'(y) \ \& \ P \ = \ P'$$

will never restore the asymmetry to a and b, since "=" is not itself asymmetrical. In ordinary terms then, the proposed analysis cannot distinguish between "aRb" and "bRa", which is of coruse crucial in asymmetric relations.

3. Prof. Ishiguro has pointed out to me that I am misrepresenting Leibniz here. I think she is surely right that the doctrine of "influx" specifically, rather than the notion of causal interaction generally is under attack at *loc. cit.* (For her reasons in some detail see Ishiguro, 243/4.) I reserve judgement however as to whether or not Leibniz has been improperly interpreted either generally or by me as holding that 'there is no connection between what happens to one thing and what happens to other things' (op. cit., p. 240), Prof. Ishiguro is masterful in her presentation of a Leibniz for whom causal interaction is a central notion, and many formerly puzzling texts are thereby clarified (see e.g. op. cit. p. 247). But there remains another Leibniz, the one who wrote to Rémond (G III, 607 = L, 655):

> Neither should you think that the monads, like points in a real space, move, push or touch one another. It suffices that the phenomena make it appear to be so and that the appearance is true insofar as those phenomena are founded.

Or again to des Bosses (G II, 444 = L, 602):

> Each [monad] is like a certain separate world, and they correspond to one another through *their*

> phenomena and have on their own account no other commerce or connection. (My emphasis.)

To yield uncritically to the physicalist view of Prof. Ishiguro is to risk introducing new puzzles for old, given the prevalence of such idealistic writings as those cited above. What is needed is a careful re-evaluation of the idealistic Leibniz in the light of the insightful discussion of his physicalist side by Ishiguro.

4. More could be said about the foundation of time in change than can be said easily here. In particular one wishes to know how exactly change can be given an atemporal model and how, in change, the prior and the posterior themselves can fail to be relational. The first question is a rather subtle one in Leibniz-exegesis and depends directly on the proper account of 'compossibility'. I have provided a limited model for it -- one whose limitations do not affect the present question -- in part V of "Compossibility, Space-Time and Possible Worlds', forthcoming in *Festschrift* for R. McRae, ed. by Profs. S. Tweyman and G. Moyal. The brief answer to the second question is that 'prior' and 'posterior' designate stages of an activity and stages presuppose the thing of which they are stages. There are convenient analogies: Except in unusual medical circumstances, it is probably not appropriate to raise questions about the *connection* between the upper and lower half of my body. The inappropriateness of this way of speaking derives from the fact that the expression 'my body', under ordinary circumstances, is taken to denote a unity which the raising of such questions would appear to challenge. Similarly inappropriate, I wish to argue, is the worry about the prior and posterior part of *one* activity. The notion of an activity involves that of a unity; an activity is not a seriate aggregation of parts.
5. I would like especially to thank Hidé Ishiguro, Calvin Normore and George Pappas for comments which led to improvements in this paper.

Robert McRae

MIRACLES AND LAWS

Leibniz makes the charge, which he constantly renewed, that the laws of nature of the Cartesians and Newton's law of gravitation were really only formulations of perpetual miracles. To make his case he had to define miracle. Because the notion of miracle involves, at least for Leibniz, the notion of law as that to which a miraculous event is an exception, and because accordingly the criteria for miracles become inversely the criteria for laws, the entire polemic throws valuable light on Leibniz's conception of law.

We may, to begin with, observe that there are three types of law for Leibniz. First, there is the law of the whole universe, sometimes also referred to as the concept of the universe. Second, there are certain architectonic principles like the law of continuity and the law of determination by maxima and minima. These laws or principles govern not only the first kind of law, the law of the universe, but also the third kind of law, the laws of nature; for example, the principle of the conservation of force in mechanics, or the principle of the most determined path in optics.

The law which governs the whole universe is the same in kind as the concept of the individual, as for example that of Julius Caesar or Alexander the Great. The concept of the individual is a law analogous to the law of a mathematical series, differing from the latter, however, in that it is a temporal series. Given the law and the starting point of the series it should be possible to deduce all the successive predicates of the subject or, if you like, deduce all the successive events in its history. Leibniz calls this kind of law a "law of order" (G IV, 518; L 493). He says, "When we say that each monad, soul, mind, has received a

K. Okruhlik and J. R. Brown (eds.), The Natural Philosophy of Leibniz, 171–181.

particular law, we must add that it is only a variation of the general law which rules the universe; and that it is just as a city appears differently according to the different points of view from which we look at it" (G IV, 553-4). The law of the individual is then a perspectival variant of the law of the universe, a law from which every event in the universe can be deduced. The law of the universe, like the laws of its individual members, is a law of order. It will appear moreover that it is logically impossible for the universe not to be subject to order. Section VI of the *Discourse on Metaphysics* is headed, "God does nothing out of order, and it is impossible even to feign events which are not regular." Leibniz explains: "Let us suppose, for example, that someone makes a number of marks on paper quite at random, as do those who practise the ridiculous art of geomancy. I say that it is possible to find a geometrical line, the notion of which is constant and uniform according to a certain rule, such that this line passes through all these points, and in the same order as the hand had marked them. And if someone drew in one stroke a line which was now straight, now circular, now of another nature, it is possible to find a notion or rule, or equation common to all the points of this line, in virtue of which these changes must occur. And there is no face, for example, the outline of which does not form part of a geometrical line and cannot be traced in one stroke by a certain movement according to rule. But when a rule is very complex what conforms to it passes for irregular." Leibniz provides another example in the *Theodicy* of the inescapable nature of order. "One may propose a succession or series of numbers perfectly irregular to all appearance, when the numbers increase and diminish variably without the emergence of any order; and yet he who knows the key to the formula, and who understands the origin and the structure of this succession of numbers, will be able to give a rule which, being properly understood, will show that the series is perfectly regular, and that it even has excellent properties" (par. 242). Up to this point we have the architectonic principle of continuity at work. Now the architectonic principle of determination by maxima and minima comes into play. Leibniz continues in the *Discourse on Metaphysics*, "Thus one can say that in whatever way God has created the world, it would always have been regular and in a certain general order. But God has chosen the one that is most perfect, that is to say the one that is at the same time the simplest in hypotheses and the richest in phenomena, as a geometrical line might be, of which the construction was easy and the properties and effects very admirable and of great extent." Or, as he puts it elsewhere "that the maximum effect should be achieved by the minimum outlay." (G VII, 303, L 487).

One may note in passing an anomalous difference in the status of the two principles. That the universe should be without discontinuities is not a matter of divine choice. Discontinuities are simply impossible. That the law of the universe should produce the most by the least is a matter of divine choice and is contingent. Elsewhere Leibniz treats the law of continuity as contingent also. Another thing we may take note of is that when Leibniz speaks of the law of the universe, the universe in question is the total aggregation of monads, spirits and minds. Let us call it the metaphysical universe in order to distinguish it from the universe of natural phenomena. The latter are governed by the laws of motion.

The next section of the *Discourse on Metaphysics*, number VII, is headed, "That miracles are in conformity with the general order, although they are counter to subordinate maxims...." He explains: "Now since nothing can be done which is not in order, one can say that miracles are as much in order as natural operations, so called [i.e. natural] because they are in conformity with subordinate maxims....As regards general or particular wills...one can say that God does everything according to his most general will, which is in conformity with the most perfect order which he has chosen; but one can also say that he has particular wills which are exceptions to the said subordinate maxims, for the most general of the laws of God which rules the whole sequence of the universe has no exceptions." Section XVII is headed "Example of a subordinate maxim or law of nature. In which it is shown that God always conserves regularly the same force but not the same quantity of motion against the Cartesians and several others." An exception to that law of nature, the conservation of force, would be a miracle. But that same miraculous event would not be an exception to the law of the universe, for exceptions to it, as we have seen, are impossible. From the law of the universe, and for that matter from the law of any individual in it, it would be possible to deduce not only those events in its history which conform to the laws of nature like the rules of collision, but also exceptions to them, that is to say, those miracles which are necessary for this to be the best of all possible worlds, or as he puts it in the *Theodicy*, (par. 248) "God ought not to make choice of another universe, since he has chosen the best and has only made use of miracles necessary thereto."

Leibniz has two related definitions of a miracle and both are continually used in his polemics. First, a miracle is that which exceeds the *power* of creatures, or that which cannot be explained by the *nature* of creatures. "Power" (or "force") and "nature" are equivalent expressions. "The nature inherent, in created things is nothing but the force to act and be acted on" (*De ipsa Natura*, par. 9). Second, a miracle is that which is not conceivable by, or explicable to, the created mind. Although miracles are included in the law of the universe, as well as in the law of each individual in it, and are perfectly intelligible to God who knows his reasons for them, finite minds are incapable of knowing the law of the universe, just as they are incapable of having the complete concept of any individual, and therefore they are incapable of understanding a miracle.

With his two definitions of, or criteria for, a miracle, Leibniz rejects absolutely any mere constant conjunction or uniformity conception of the laws of nature, such as he found in the Cartesians and in Bayle defending them, and also in Clarke's defence of Newton. The issue arises first in connection with the Cartesians' occasionalist account of the relation of mind and body, but it is then almost immediately extended to the occasionalist account of all secondary causes. Bayle in his article on Rorarius remarked, "Furthermore it seems to me that this able man [Leibniz] dislikes the Cartesian system because of a false assumption, for one cannot say that the system of occasional causes makes the action of God intervene by a miracle (*deus ex machina*) in the reciprocal dependence of body and soul. For since God's intervention follows only general laws, he does not therein act in an extraordinary way." In replying to this Leibniz after discussing the mind and body connection goes on to say, "But let us see whether the system of occasional causes does not in fact imply a perpetual miracle. Here it is said that it does not, because God would act only through general laws according to this system. I agree, but in my opinion that does not suffice to remove the miracles. Even if God should do this continuously, they would not cease being miracles, if we take the term not in the popular sense of a rare or wonderful thing, but in the philosophical sense of that which exceeds the powers of created beings. It is not enough to say that God has made a general law, for besides the decree there is also necessary a natural means of carrying it out, that is all that happens must be explained through the nature which God gives to things. The laws of nature are not so arbitrary and indifferent as many people imagine. For example, if God were to decree that all bodies should have a tendency to move in circles and that the radii of the circles should be proportional to the magnitude of the bodies, one would either have to say that there is a method of carrying this out by means of simpler laws, or one would surely have to admit that God must carry it out miraculously, or at least through angels charged expressly with this responsibility....It would be the same if someone said that God had given natural and primitive gravities to bodies by which each tends to the centre of its globe without being pushed by another body, for in my opinion such a system would need a perpetual miracle, or at least the help of angels." (G IV, 520-1; L 494-5). To Arnauld Leibniz writes; "Strictly speaking God performs a miracle whenever he does something that exceeds the forces which he has given to creatures and maintains in them. For instance, if God were to cause a body which had been set in a circular movement, by means of a sling, to continue to move freely in a circle when it had been released from the sling, without being impelled or checked by anything at all, that would be a miracle, for according to the laws of nature it should continue along in a straight line in a tangent; and if God were to decree that that should always occur, he would be performing natural miracles, since this movement is not susceptible of a similar explanation. Likewise one must say that if the continuation of the movement exceeds the force of the bodies, it must be said according to the accepted concept, that the continuation of the movement is a true miracle, whereas I believe that bodily substance has the force to continue its changes according to the laws that God has placed in its nature and maintains there." (G II, 93) Or again, in commenting on the occasionalist Lami's objection to his system: "If God wills that a body tend of itself in a straight line, that will be a law of nature, but if he wills that of itself it goes in a circular or elliptical line that will be a continual miracle." (Rob. 373).

The second criterion of the miraculous is that it is inconceivable to created minds. Thus "A miracle is a divine action which transcends human knowledge; or more strictly which transcends the knowledge of creatures" (C 508). By conceivable or intelligible Leibniz means quite simply that and that only which is susceptible of mechanical explanation. To Lady Masham he writes, "It is well to consider that the ways of God are of two kinds, the ones natural, the others extraordinary or miraculous. Those which are natural are always such as a created mind would be able to conceive...but the miraculous ways lie beyond any created mind. Thus the operation of the magnet is natural, being entirely mechanical or explicable, although we are still perhaps not in the position of explaining it perfectly in detail, for want of information; but if anyone maintains that the magnet does not operate mechanically and that it does it all by pure attraction from a distance, without any intermediary, and without visible or invisible instruments, that would be something inexplicable to any created mind, however penetrating and informed it should be; and in a word it would be something miraculous," (G III, 353) or again in the *New Essays* he says "Matter cannot naturally attract...nor of itself proceed in a curved line, because this cannot be explained mechanically, for that which is natural must be capable of being distinctly conceived" (Pref.) Not even God could get us to understand a non-mechanical explanation, as Leibniz points out in a letter to Hartsoeker. "Thus the ancients and moderns, who avow that weight is an occult quality are right if they mean by that that there is a certain mechanism unknown to them, by which bodies are pushed towards the centre of the earth. But if it is their opinion that the thing is done without any mechanism by a simple primitive quality or by a law of God which produces this effect without using any intelligible means, it is an irrational occult quality which is so occult that it is impossible that it can ever become clear, even if an angel, to say nothing of God himself, wanted to explain it" (G III, 519).

With Newton and his spokesman, Clarke, Leibniz came up against a rival conception of the natural and the miraculous, not the less forcefully stated because of the insulting tone in which Leibniz in the opening letter of the exchange with the Englishmen makes the charge that Newton's God has created a world so imperfect that he must resort to miracles to keep the whole machine going, while Leibniz's God with the law of the conservation of force has no need of any such extraordinary interventions. "And I hold," says Leibniz, "that when God works miracles, he does not do so in order to supply the wants of nature, but of grace." (Leibniz, I, 4). From this point on miracles are a major topic throughout the correspondence generating, perhaps, the most heated mutual scorn in the entire exchange.

Clarke's conception of a miracle comes out first in his second letter in which he maintains that the distinction between the natural and the supernatural has no significance in relation to God, but only to human ways of conceiving things. For God nothing is more miraculous than anything else. But for us the natural is the usual or frequent or regular, the supernatural the unusual. Says Clarke, "The raising of a human body out of the dust of the earth, we call a miracle; the generation of a human body in the ordinary way we call natural; for no other reason but because the power of God affects one usually, the other unusually. The sudden stopping of the sun (or earth,) we call a miracle; the continual motion of the sun (or earth,) we call natural; for the very same reason only, of the one's being usual and the other unusual. Did a man usually arise out of the grave, as corn grows out of seed sown, we should certainly call that also natural; and did the sun (or earth,) constantly stand still, we should then think that to be natural, and its motion at any time would be miraculous" (Clarke, V, 107-109). It is evident that for Clarke it is not the kind of causes involved which determines what is natural or supernatural. "The means by which two bodies attract one another, may be invisible and intangible, and of a different nature from mechanism; and yet, acting regularly and constantly, may well be called natural." (Clarke, IV, 45). To which Leibniz replies: "He might as well have added inexplicable, unintelligible, precarious, groundless and unexampled" (Leibniz, V, 120). Growing tired of being told repeatedly that the natural is that which can be explained by the natures of creatures, Clarke finally protests that "The terms, *nature*, and *powers of nature*, and *course of nature* and the like are empty words; and signify merely that a thing usually or frequently comes about." (Clarke, V, 107-109). As for Leibniz's calling attraction a miracle or an occult quality Clarke finds this most unreasonable "after it has so often been declared," - and here Clarke quotes several of Newton's disclaimers, which he assumes that Leibniz will have read - "that by that term we do not mean to express the cause of bodies tending towards each other, but barely the effect, or phenomenon itself, and the laws or proportions of that tendency discovered by experience; whatever be or be not the cause of it." (Clarke, V, 110-116). If Leibniz had lived long enough to read these remarks the dispute should have moved away from miracles to the question of the relation of laws and causes, but again there would have been no meeting of minds.

I want now to consider Leibniz's conception of miracles and laws in relation to his celebrated doctrine of possible worlds of which this world is one. To Arnauld (G II, 40) he says, "If this world were only possible, the individual concept of a body in this world, containing certain movements as possibilities, would also contain our laws of motion (which are free decress of God) but also as mere possibilities. For as there exists an infinite number of possible worlds, there exists also an infinite number of laws, some peculiar to one world, some to another, and each possible individual of any one world contains in the concept of him the laws of his world." It should be noted that usually when Leibniz refers to possible worlds, it is to the metaphysical worlds of monads, spirits, minds. In this letter to Arnauld the possible worlds are phenomenal worlds of bodies, not monads or minds.

Leibniz has effectively excluded as laws of any possible world such uniformities as the tendency of all bodies of themselves to move in circles, or to tend to the centre of their globes without being pushed there by other bodies, to cite examples with which he responds to Bayle. These would be miracles in all possible worlds according to Leibniz because (a) they are not consequences of the natures which God gives things and (b) they are not conceivable or explicable or intelligible, and these are the criteria of the natural as opposed to the supernatural. Are we then left with any other possible laws of nature for other worlds than those operating in this world? To begin with, can God give bodies natures other than that *vis viva* which he conserves in this world? It would appear not. Leibniz rejects as occult and unintelligible any other candidates for the nature of a body. In any case if we are talking as Leibniz does to Arnauld about the bodies in different possible worlds, these will all, if they are to be bodies, by definition share the same nature. Given, then, that all possible bodies have this nature, and that only mechanical explanations are intelligible, can there be laws of motion which are possible alternatives for other worlds to those operative in this world? What Leibniz calls "the most universal and inviolable" (G III, 45) law of nature, i.e. the conservation of force, he also regards as "the foundation of the laws of motion," (*De ipsa Natura* par. 4) and indeed claims to "reduce all mechanics" to it (G II, 62). What we are asking, then, is whether there is such a foundational law, other than the conservation of force, peculiar to each of an infinitude of possible worlds and to which all the laws of motion peculiar to each of those worlds can be reduced? The answer would seem clearly to be no, for Leibniz says "I call extraordinary every operation of God demanding something other than the conservation of the nature of things." (A 185). Since the nature of bodies is the same in all possible worlds, i.e. force, an exception to the law of the conservation of force would be a miracle in all possible worlds. It would appear then that Leibniz's conception of the miraculous commits him, contrary, of course, to his deepest intentions, to holding that the laws of motion in this world are the same for all possible worlds, the laws, that is to say, of Leibnizian mechanics. Not only would Euclid's *Elements* be a text-book in all possible worlds, but so also would the elements of Leibnizian mechanics.

As an addendum I should like briefly to consider the relation to possible worlds of another set of laws, those of optics, without reference, however, to the miraculous, but to Leibniz's use of the two architectonic principles, that of order or continuity and that of determination by maxima and minima. In the *Tentamen Anagogicum* he combines them to produce the principle of the most determined or unique path for light rays in order to find the laws of reflection or refraction. The unique path is that which has no twin or other path symmetrical with it. All other paths have twins. Implicit in Leibniz's use of this principle is another which he enuciates in his correspondence with Clarke, namely "When two things which cannot both be together, are equally good; and neither by themselves, nor by their combination with other things, has the one any advantage over the other; God will produce neither of them." (Leibniz, IV, 19). God is helpless in choosing between twins; hence if he is to choose at all he must choose the most determined or unique path. The concept of unique determination is purely spatial. If, then, there can be different laws of optics for different worlds, it can only be a consequence of the possibility of different kinds of space for these worlds. Leibniz denies this possibility. "Why," asks Bayle, "has matter precisely three dimensions? Why should not two have sufficed for it? Why has it not four?" To which Leibniz replies, "the ternary number is not determined by reason of the best but by geometrical necessity because geometers have been able to prove that only three straight lines perpendicular to one another can intersect at one and the same point." (*Theodicy*, par. 355). It looks then, as if the laws of optics will join those of motion as applying in all worlds, with, however, this possible qualification. Leibniz indicates to Arnauld that there are bodies and motion in all worlds. Does he believe that there is light in all worlds? By the following speculative steps we must, I think, come to the conclusion that he does. Leibniz says, "For by the individual concept of Adam I mean, to be sure, a perfect representation of a particular Adam who has particular individual conditions and who is thereby distinguished from an infinite number of other possible persons who are very similar but yet different from him (as every ellipse is different from the circle, however much it approximates to it" (G II, 20). In other words the existing Adam is the last term upon which an infinite series of possible Adams converges in the same way as an infinite series of ellipses converges on the circle. If the law of the individual is only a perspectival variant of the law of the universe, there must be a similar series of possible worlds converging on this world. If that is the case for the metaphysical worlds of individuals or monads, it is, perhaps, not too much to suppose that it is the case also for its phenomenal counterpart, the physical world, and consequently there should be ever increasing degrees of illumination in the series of possible worlds converging on this world. If so, Snell's law of refraction must apply in all possible worlds.

Department of Philosophy
University of Toronto

References

A *The Leibniz-Clarke Correspondence*, ed. by H.G. Alexander, Manchester, 1956.

C *Opuscules et fragments inedits de Leibniz*, ed. by L. Couturat, Paris, 1903.

G *Die philosphischen Schriften von G.W. Leibniz*, ed. by C.I. Gerhardt, 7 vols, Berlin, 1875-90.

L *Philosophical Papers and Letters*, trans. by L.E. Loemker, Dordrecht, 1969.

Kathleen Okruhlik

THE STATUS OF SCIENTIFIC LAWS IN THE LEIBNIZIAN SYSTEM

The aim of this essay is to secure as painless and as firm a grasp as possible on the very thorny problem of the status of scientific laws in the Leibnizian system. The thorniness of the problem is immediately evident when we realize that Leibniz maintains that laws of nature are absolutely contingent, hypothetically necessary, and (in some cases at least) *a priori* deducible.

Leibniz was driven to this complex doctrine partly in an effort to reconcile the explanatory scope of mechanics with the preservation of two divine attributes: God's perfect rationality, on the one hand, and his perfect freedom on the other. Very closely related to this second attribute is God's providential concern for his creatures, a property which Leibniz thought was manifested most clearly in the laws of nature.

To understand why Leibniz chose to steer the tortuous path that he did regarding the metaphysical and epistemological status of scientific laws, it is necessary to appreciate the dangers he perceived in alternative accounts. To this end, I shall sketch very briefly and in a somewhat oversimplified fashion a picture of each of three rival views which Leibniz found dangerous: those of Spinoza, Newton, and Descartes.

The Historical Background

According to Spinoza, nothing in the universe is really contingent. Individual existing things are modes of the one necessary substance whose essence entails existence. Thus, everything which exists does so from the necessity of God's essence; and everything which happens does so because

K. Okruhlik and J. R. Brown (eds.), The Natural Philosophy of Leibniz, 183–206.

of the laws of God's nature. Nothing could have been other than as it is:[1]

> Things could not have been brought into being by God in any manner or in any order different from that which has in fact obtained.

This result is repugnant to Leibniz because it robs God of his proper freedom. Spinoza maintains that God is free in the sense that his actions are determined, not by any external influence, but by his nature alone. This is not good enough for Leibniz. If God were not free to choose among possible worlds in deciding which to create, then there would be no room in the world for design or for the exercise of divine providence. Yet there is evidence of design everywhere, according to Leibniz. So Spinoza's account is not only theologically suspect, it is empircally inadequate.

Leibniz cannot fault Spinoza on the question of rationality; Spinoza's system is highly rational. Every connection is a logical connection. To explain why some object possesses the attributes that it does or why it undergoes certain modifications is to exhibit these attributes or modifications as the strictly logical consequences of other attributes or modifications. Leibniz's complaint is that Spinoza purchases this perfect rationality at too high a price: By grounding existence and the laws which govern existing things so firmly in God's essence, Spinoza preserves rationality by eliminating contingency.

The Newtonian philosophy suffers from the opposite defect in the eyes of Leibniz. On this view, there is too much room for divine whimsy and too little room for rationality. God does whatever pleases his fancy, with or without reason. Or rather, God's will in itself constitutes sufficient reason for any act, however lacking in rationality it might be otherwise. Samuel Clarke represents the Newtonian position in these words:[2]

> 'Tis very true that nothing is without a sufficient reason why it is and why it is thus rather than otherwise....But this sufficient reason is ofttimes no other than the mere will of God....Which if it could in no case act without a predetermining cause, any more than a balance can move without a preponderating weight, this would tend to take away all power of choosing and to introduce fatality.

Leibniz replies[3] that "a mere will without any motive is a fiction, not only contrary to God's perfection but also chimerical and contradictory."

Clarke does not budge from his position because he believes that to prohibit God from acting without a good reason is to unduly diminish his freedom. This so-called "freedom of arbitrariness," however, is to Leibniz the very antithesis of genuine freedom and a product of sacrilege.

If we maintain that God can act in the absence of a reason, then we not only detract from his rationality, we also undermine one of the epistemological bases of natural science. If we leave open the possibility that aspects of the existing order of nature are without rational foundation, then it follows that one set of scientific laws is just as likely as any other; and we cannot find *a priori* foundations for those which actually obtain. All knowledge of scientific laws on this account is strictly and irremediably *a posteriori*. This limitation is not a merely epistemological limitation owing to the nature of the human knower; rather it reflects the character of the universe itself. It is possible that some laws of nature are entirely without rational foundation and that their obtaining is the result of a raw exercise of divine power. If this were true, then the intelligibility of the universe would be radically limited.

Leibniz certainly concurred in the belief that most of our knowledge of nature must be obtained *a posteriori*. And with increasing age, he appears to have grown progressively less sanguine about the ability of human knowers to actually grasp the *a priori* reasons for specific scientific laws. He steadfastly maintained, however, that -- quite apart from human ability to grasp it -- there must exist in the essence of things some reason for their existence and some reason for the laws which govern their interactions. The Newtonians were able to avoid Spinozan necessitarianism only by denying this. The price they paid for maintaining the contingent status of scientific laws was a radical restriction in divine rationality and hence in the ultimate intelligibility of nature. Spinoza had insisted on a necessary linkage between the realm of essences and the existing order of nature, thereby preserving rationality and eliminating contingency. The Newtonians, on the other hand, severed the necessary linkage between essence and the existing order, thereby preserving contingency while limiting rationality.

We turn now to the third account of the status of scientific laws which Leibniz strove to avoid -- that of Descartes. Cartesian mechanics is not so easy to characterize on the essence-existence model employed above. In the final analysis, it comes out looking much like the Spinozan account, though not so extreme. As Margaret Wilson has pointed out[4], it is rather difficult to understand just what Leibniz means when he accuses Descartes of assigning to the laws of nature the status of geometric necessity. At first blush, the very opposite may appear to be true. Descartes's God, unlike the God of Leibniz, controls even the realm of essences. Relationships among essences are not for him (as for Leibniz) unassailable matters of logic beyond God's power to affect. Rather essences themselves are determined by God. This seems to be a far cry from necessitarianism. Yet Cartesian mechanics was viewed by Leibniz and others as a sort of necessitarian system. Why should this be so?

The thinking seems to be that *once God decided* that the essence of matter should be extension, he thereby established a set of laws which govern the interactions of material bodies *in all possible worlds.* The argument goes something like this: If body is material and if the one essential property of matter is extension (as Descartes maintained), then all the laws governing the behavior of bodies must derive from this one property of extension. But extension is a purely geometrical property, and the laws governing extension are geometrical laws. The laws of geometry hold in all possible worlds. Therefore, the laws governing the behavior of matter hold in all possible worlds. (It doesn't follow from this that there is only one possible world. The range of initial conditions is infinite, and God was not constrained in his arrangement of these conditions.)

This argument may not be entirely just to Descartes, but it does fairly characterize what his opponents perceived as the geometric spirit of Cartesian mechanics. We can deduce laws of mechanics *a priori* because the same laws must obtain in any world in which bodies exist. Descartes himself says that the laws which he deduces in his physics "are such that even if God had created several worlds, there could be none in which these laws failed to be observed."[5]

This, though not Spinozistic, is too necessitarian for Leibniz. He wants to maintain that the laws governing the behavior of bodies *could* have been different and that there are possible worlds in which they *do* differ. This allows room for the role of final causes in God's choice among laws.

Leibniz's task, therefore, is to tell a story which makes the existence of this world and of its laws rational without thereby making it necessary.

The laws of nature must not follow deductively from the essence of matter or from the essence of God. On the other hand, there must be a good reason certain essences were realized while others were not. These restrictions set the stage for Leibniz's account of creation. In detailing that account, I shall rely in part on E.M. Curley's very helpful exposition in an article entitled "The Roots of Contingency".[6]

Compossibility, Creation, and Contingency

The first thing we must be clear about is the fact that Leibniz does not believe that all possibles exist. He maintains that between any two existing things there are infinitely many other existents, but that entire series of possible existents remain unactualized:[7]

> I have reasons for believing that not all possible species are compossible in the universe, great as it is, and that this holds not only in respect to the things which exist together at one time, but even in relation to the whole series of things. That is, I believe that there necessarily are species which never have existed and never will exist, not being compatible with that series of creatures which God has chosen... The law of continuity states that Nature leaves no gap in the order which she follows; but not every form or species belongs to every order.

It is evident then that Leibniz has a more stringent criterion of existence for substance than mere lack of internal contradiction. Two substances realizable in themselves may not be jointly realizable, and we are, therefore, forced to require that every existent be *compossible* with every other existing thing.

Now a problem arises here in determining just how two substances could fail to be compossible, given Leibniz's account of substance. It is well-known that he believed all purely positive predicates to be entirely independent of one another, thus seeming to preclude the possibility of conflict among substances. Leibniz himself recognizes the difficulty:[8]

> It is not yet known to men from what the incompossibility of different things arises or how it comes about that different essences are opposed to one another, since all purely positive terms appear to be compatible *inter se*.

Russell discussed this problem in 1900 and offered a possible solution. He suggested[9] that perhaps only those things are compossible which can co-exist in a *cosmos*; that is, in a universe subject to the reign of law. The reasoning goes something like this: Nothing is done except to achieve an end. Different laws facilitate the achievement of different ends; and where there is no law, no end whatsoever can be achieved. Thus, a chaotic collection of substances can serve no purpose. From this it follows that there could never be a sufficient reason for bringing into existence such a collection. Though such a world could not be excluded on the grounds of geometrical necessity (that is to say, it does not contain within itself either an explicit or implicit contradiction), it could be ruled out by a sort of metaphysical necessity based on the principle of sufficient reason. Now there is very slim textual evidence for this position,[10] and there is even evidence against it.[11] Leibniz seems to indicate in places that God *could* have created a chaos had he wanted to.

It would appear that Russell's thesis is too strong. Compossibility does not require law-governedness, but something weaker. This "something weaker" is the pre-established harmony among the fundamental units of being, the monads. Recall that Leibniz's monads are "windowless"; there is no interaction among them. Nevertheless, God can arrange things in such a way that monads apparently "reflect" one another (where this reflection is entirely non-causal). Let us say, for the sake of a simple example, that you and I are the sole occupants of a two-monad universe. We are each completely programmed from the moment of creation. (The program metaphor is Rescher's.) Everything that will ever happen to me is written into my program, and everything that will ever happen to you is written into yours.

It may be the case that our two programs are co-ordinated with one another so that my perception of you shaking my hand is co-ordinated with your perception of my shaking your hand. There would appear to be no logical contradiction, however, in supposing that our programs are entirely unco-ordinated with one another. We could be like two people sleeping side by side having two different dreams. Perhaps you don't even figure in my dream. Perhaps you do; but while I dream that we are scaling Mt. Everest, you dream that we are solving fundamental problems in metaphysics. There would appear to be no logical contradiction in supposing that such a universe could exist.

It seems plausible to maintain that in the first case the monads are compossibles while in the second case they are not. In neither instance would there be any strictly logical contradiction in maintaining that the two monads both exist. In the latter case, however, there is an important sense in which the two monads cannot exist *together*; they cannot form a *community* of any sort. So we call them "incompossibles".

Groups of compossibles, therefore, are groups of monads whose programs are mutually co-ordinated in such a way as to give rise to the appearance of interaction. Leibniz maintains that this co-ordination is *complete*: among a group of compossibles, every monad reflects *every* predicate of *every* other monad. An immediate consequence is that no monad can belong to more than one group of compossibles.

Notice that a world of compossibles is not necessarily a *law-governed* world in any usual (non-trivial) sense of that word. Every monad may appear to be interacting with every other monad, but this interaction may be chaotic. It is on this basis that it was alleged earlier that Russell's criterion for incompossibility was too strong. If we regard each group of compossibles as a possible world, then the *law-governed* possible worlds form a proper subset of the set of all possible worlds. So even if we limit God's choice to groups of compossibles, we do not preclude him from creating a world which is not strictly law-governed.

The overall picture is this: Compossibility is an equivalence relation. We start with the set of all possible existents. We use the compossibility criterion to partition this set into mutually disjoint subsets. Each subset is an equivalence class; i.e., every element belongs to *exactly one* such subset. Now consider the set of all these equivalence classes. This is the set of all possible worlds. The set of all *law-governed* possible worlds is a subset of it. This existing world is, of course, an element of that subset.

Notice that treating compossibility as an equivalence relation points to a very important difference between Leibniz's theory of possible worlds and the approach taken by many contemporary possible world theorists. These latter are willing (and anxious) to talk about a possible world in which Caesar did not cross the Rubicon. On Leibniz's account, however, this is nonsense. There is no possible world in which there is Caesar but he does not cross the Rubicon. It is an essential property of Caesar that he crosses the Rubicon. So a person very much like Caesar except that he did not have this one river-crossing property would *not* be Caesar. Caesar is to be found in only one possible world. This is what it means to say that compossibility is an equivalence relation.

It is now necessary to show how this notion of compossibility functioned in Leibniz's attempt to steer a middle course between Newton, on the one hand, and Descartes and Spinoza, on the other.

What we have established so far is that the class of candidates for existence comprises all those collections of things compatible with one another. The next question facing us is the following: On what basis did one of those collections become actualized rather than any other collection? What we are asking, therefore, is why this particular world exists rather than any other. The answer seems to be that this world realizes a greater quantity of essence than could any other possible world; it allows for the existence of more *being* than could any other.[12]

> I say, therefore, that the existent is the being which is compatible with most things, or the most possible being so that all co-existent things are equally possible.

Having answered one question, we are immediately confronted with two others. First, is the realization of maximum essence necessary or contingent? That is to ask, are worlds in which being is *not* maximized metaphysically allowable entities? Secondly, is there room in the Leibnizian account for God and for a Christian type of creation? Or are essences self-realizing, thus rendering God superfluous? The classic statement of the view that essences are, in fact, self-actualizing was provided by Russell[13] in 1903 after he read Couturat:

> Essences range themselves in the conflict on the side of those with which they are compossible and a tug of war results in which the majority are victorious. An interesting conflict of ghosts all hoping to become real! But it is hard to see what God has to do in that *galère.*

Before reading Couturat[14], Russell had believed that what exists does so contingently and by virtue of a free decree on God's part. He changed his mind on the basis of new textual evidence unearthed by Couturat to the effect that existence is for Leibniz a predicate. This would mean that existential propositions whose subject referred to an existing entity would be *necessarily* true by virtue of being of the form "AB is B".

What we must first establish is whether or not Leibniz actually did believe that existence was a predicate. There are passages where he certainly seems to say so:[15]

> Existence is commonly conceived by us if it were a thing having nothing in common with essence, which nevertheless cannot be the case, because there must be more in the concept of the existent than in that of the non-existent, i.e., existence is a perfection, since there is really nothing else explicable in existence than that it enters into the most perfect series of things.

In his work *On the Radical Origination of Things,*[16] Leibniz argues that the very fact that something exists rather than nothing must convince us that

> ...there is in possible things, or in possibility or essence itself, some exigency of existence, or so to speak, a reaching out for existence, or in a word, that essence of itself tends toward existence. Whence it follows that all possibles, i.e., things expressing essence or possible reality, tend by equal right toward existence, according to the quantity of essence or reality, or the degree of perfection which they involve.

In another place,[17] a case is made to the effect that if Adam a sinner exists, then there must be in the essence of Adam something by virtue of which he does exist.

Finally, there is a passage[18] in which Leibniz formulates a sort of *reductio* of the position that existence is something superadded to essence. If it were, he argues, then that something would have an essence of its own; and we would have to ask in virtue of what *it* exists...and so on *ad infinitum*.

Not only does there seem to be direct textual support for the position that Leibniz believed existence to be a predicate, there is also strong indirect evidence in the fact that he advanced a form of the ontological argument. To maintain, of course, that such an argument proves the existence of God, one must assume that existence is a perfection, i.e., a positive predicate.

On the other hand, Leibniz explicitly denies in other places that the essence of a substance entails its existence, as when he says that "The possibility or notion of a created mind does not involve existence."[19] How then do we resolve this dilemma? Is existence a part of the essence of substances, or is it something conferred upon essences by God? Leibniz clearly believed that essences were objects of God's intellect, but that existence was a product of his will. The Russellian tug of war occurs only within the mind of God, and the victorious essences are absolutely dependent upon God for actualization. But if essences are not self-actualizing, then how can we explain Leibniz's assertions that existence is an exigency of essence? The best answer is found in Grua 288:[20]

> All truths about contingents, i.e., about the existence of things, depend on the principle of perfection. All existences, the existence of God alone excepted, are contingent. But the reason why one contingent thing exists rather than another, is not sought from its definition alone, but from comparison with

> other things. For since there are infinitely many possibles which nevertheless do not exist, the reason why these exist rather than those must not be sought from the definition (otherwise, not to exist would imply a contradiction, and the others would not be possible, contrary to the hypothesis), but from an extrinsic principle, which is that these are more perfect than the others.

There are two important things to note about this passage. First, it clearly denies that any *individual* essence possesses an absolute claim to existence. Read in conjunction with the text referred to in footnote 15, this passage indicates that any claim to existence (whatever its nature) is associated with whole *series* of compossible essences (i.e., equivalence classes) on a comparative basis, not with individual essences on an absolute basis. Secondly and more importantly, the passage makes clear the importance of the principle of perfection. The claim to existence possessed even by a series of substances *presupposes* that God has decided to maximize essence. All pretensions of this world to existence depend absolutely on the implementation of this principle. Clearly then, even a series of essences is powerless to actualize itself; it exists only as an object of God's thought until he wills to make it actual. Now once God has decided to employ the principle of perfection (i.e, to maximize entity because entity is good), he like everyone else is bound by the laws of logic. And the creation of *this* world follows inexorably from that decision.

We have established, then, that there definitely is a role for God in the Leibnizian account of existence. We have also established a case for the *hypothetical* necessity of the existence of this world. It is necessarily the case that *if* God decides to implement the principle of perfection, *then* this world exists. A larger question remains, however; it is the one posed earlier in this section. Is the maximization of essence necessary or contingent? That is to ask, did God have a choice? A strong case can be made to the effect that he did not, based on the fact that goodness is an essential attribute of God. Leibniz emphatically denies that this is the case. God's intellect, he says, is bound by the laws of logic; his will is entirely free. Though considerations of goodness *inclined* him to actualize a maximum of being, they did not necessitate his decision.[21]

> I think you will concede that not everything possible exists....But when this is admitted, it follows that it was not from absolute necessity, but from some other reason (as good,

order, perfection) that some possibles exist rather than others.

And again:[22]

> When anyone has chosen in one way, it would not imply a contradiction if he had chosen otherwise, because the determining reasons do not necessitate the action.

The first thing to note here is that God's choice of this world was anything but arbitrary. Arbitrariness, as we have seen, is for Leibniz the very antithesis of freedom. The freest choice is the choice most informed by reason.

We can assent to Leibniz's account of freedom and say that God acted freely in creating this universe because the principle of perfection gave him good reason to do so. We would then have to ask, however, just why God decided to adopt the principle of perfection. Either he did it arbitrarily or he did it for a reason. He could not have done it arbitrarily because nothing acts arbitrarily in Leibniz's system, least of all God. But if he did it for a reason, then this reason must involve the choice of some prior end. This prior end was chosen either arbitrarily or for a reason...and so *ad infinitum*. To most readers this would appear to be an effective *reductio ad absurdum*. Leibniz, however, happily embraces the infinite regress, seeing nothing in it which is beyond God's powers.[23]

> If anyone asks me why God has decided to create Adam, I say, because he has decided to do the most perfect thing. If you ask me why he has decided to do the most perfect thing or why he wills the most perfect...I reply that he has willed it freely, i.e., because he willed to. So he willed because he willed to will, and so on to infinity.

This answer will satisfy hardly anyone; and it must be admitted that this is one of the weakest points in Leibniz's metaphysics. Many critics maintain that Leibniz's efforts to ward off Spinozism break down completely at this point. They claim that the contingency of the existence of this world and hence the contingency of the laws of nature are hopelessly compromised because (in the final analysis) God had no choice but to create this world.

As Margaret Wilson has pointed out, however, it is important to note that even if the critics are right on this point, very significant differences remain between the Leibnizian system and the necessitarianism of Spinoza. As she says, "the main point can be expressed very simply:

Leibniz's philosophy requires that the explanation of any existential proposition involve reference to value, purpose, perfection."[24] Futhermore, it should be pointed out even if God *had* to create this world, it does *not* follow that the connections among phenomena which are expressed in scientific laws must be logical/conceptual connections as they are in Spinoza.

In any event, it is clear that on this account, the contingency of the laws of nature stands or falls with God's freedom to accept or reject the principle of perfection. Leibniz tried to steer a middle course between Spinoza's and Descartes's necessitarianism and Newton's arbitrariness by arguing that there is a reason in the essence of each existing thing that it should exist, but that this reason does not *entail* existence.

Contingency and Infinite Analyzability: Are the Laws of Nature Analytic?

There is a second (perhaps supplementary) account of contingency which is common in the literature. It maintains that the difference between necessary and contingent propositions is to be located in the length of the analysis required in order to reveal that the predicate inheres in the subject. There certainly are passages in which Leibniz appears to maintain that all true propositions, both necessary and contingent, are analytic in form. That is to say, the predicate of any true proposition can through proper analysis be found to be contained in the subject. The only difference between necessary and contingent truths on this account lies in the length of the analysis. In the latter case, it is infinitely long and can be successfully achieved only by God. Leibniz employs one of his favorite analogies to make this point in the following passage:[25]

> It is essential to discriminate between necessary or eternal truths, and contingent truths or truths of fact; and these differ from each other almost as rational numbers and surds. For necessary truths can be resolved into such as are identical, as commensurable quantities can be brought to a common measure; but in contingent truths as in surd numbers, the resolution proceeds to infinity without ever terminating. And thus the certainty and the perfect reason of contingent truths is known to God only, who embraces the infinite in one intuition. And when this secret is known, the difficulty as to the absolute necessity of all things is removed, and it appears what the

difference is between the infallible and the necessary.

It is clear from other passages that contingent propositions, or as Leibniz often calls them "propositions of fact", are those which in some way involve existence.

The idea behind the thesis that contingency arises from infinite analyzability is something like the following: Every existing thing is a concrete individual; i.e., all of its predicates are fully specified. Because there are an infinite number of substances and because each substance reflects every other substance, the number of these predicates is infinite. Furthermore, it follows that each individual substance brings with it an entire universe; as soon as you have one individual, you have a fully specified possible world. Another way of saying this is by pointing out that each individual substance picks out an equivalence class of compossibles. Therefore, to analytically demonstrate the truth of a proposition about an individual, one must know everything about the world of which it is a part. Clearly, this task is impossible for human beings, possible for the deity. Hence, contingency is rooted in infinite analyzability. The distinction between necessary and contingent propositions appears on this account to be more epistemological than ontological.

Leaving aside for a moment the question of whether or not Leibniz himself adhered to this position, let us examine it to see whether (in itself) it constitutes a plausible explanation of contingency. The argument here will be to the effect that it does not.

Consider the sentence "Caesar crosses the Rubicon". There are two ways of interpreting this sentence. One is purely intensional. It makes no reference to actually existing substances but asserts that the *concept* "Caesar" contains the *concept* "crosses the Rubicon". According to the second interpretation, this sentence says that the substance named Caesar *exists* (i.e., is a member of this possible world) and that he crosses the Rubicon. On the first interpretation, the sentence has no existential import and only expresses a relationship of ideas. Therefore, we would expect it to be a necessary truth. (Recall that relationships among essences are unassailable matters of logic which even God cannot change.)

On the second interpretation, however, the sentence does have existential import; it expresses a matter of fact. Therefore, we would expect it to be contingent.

Notice, however, that with respect to the complexity of the subject term, the two interpretations are on a par. If we maintain that "Caesar" is an infinitely complex subject term, then on the infinite analyzability criterion of contingency, both interpretations should come out contingent. This is clearly wrong.

The infinite analyzability criterion does *not* separate the contingent from the necessary propositions. At best it separates those propositions whose subject term names a concrete individual from those whose subject term names something else. This consideration suggests a very powerful objection to the view of contingency as based on the infinite analyzability of the subject term: This view would appear to make the laws of nature come out necessary. Yet this is precisely the consequence Leibniz was trying to avoid. He stresses repeatedly that scientific laws are paradigmatic instances of contingent propositions:[26]

> These laws are not necessary and essential but contingent and existential....For since it is contingent and depends on the free decree of God that the series of things exists, the laws of the series will also be contingent.

The argument goes something like this: If contingency were grounded in infinite analyzability of the subject term, then it would be applicable only to *complete* notions; i.e., to individual substances. Laws of nature, however, involve abstraction, idealization, and generalization. They employ "full" rather than "complete" concepts, to use Leibniz's terminology. Therefore, on the view of contingency which is being argued against here, the laws of nature could not possibly be contingent.

This requires unpacking. Recall that the complete notion of an individual substance entails everything that will ever happen to it and that the individual substance reflects everything that will ever happen in its own world. Thus the complete notion of an individual substance brings with it, so to speak, an entire universe and renders infinite the analysis of any proposition the subject of which is an individual existent.

The case is different with respect to abstractions and generalizations. Consider a species, for example. We might want to argue that it is possible to know all the defining characteristics of a species. At the same time, however, this information does not tell us anything about the vicissitudes of the lives of individual members of that species. The case is the same with terms like "kingship". We can know everything there is to know about kingship without knowing everything about Alexander. Thus, Leibniz refuses to call abstract concepts "complete", however exact they may be. Instead he reserves for these the term "full".[27]

Now what bearing does all this have on the status of physical laws? It should be fairly obvious. A law is not a law unless it abstracts to some degree from concrete individuals. The defining characteristic of laws and the source of their usefulness is to be found in generalization and abstraction. They enable us to deal with *classes* of events and *classes* of objects -- e.g., to treat individual bodies *qua* massive or *qua* moving without regard to color. The *raison d'etre* of laws is precisely that they allow us to *escape* the infinite complexity of individual substances. If it is this infintite complexity which grounds contingency, as the view we are opposing maintains, then laws of nature clearly cannot be contingent. But they are. Therefore, the view which locates the roots of contingency in the infinite analyzability of the subject term must be mistaken.

What if we maintain that infinite analyzability refers not to the infinite complexity of the subject term but to the infinite process which God undertook at the time of creation of comparing all possible worlds to one another? Can this sort of "infinite analyzability" be the hallmark and source of contingency? The answer, I think, is that this type of infinite analysis may be a proper *sign* or *mark* of contingency but that it cannot be the source of contingency. Notice first that this brand of infinite analysis does succeed in picking out contingent propositions -- i.e., those with existential import. If a proposition makes no existence claim, then it is irrelevant whether or not its subject term is a member of the best possible world; and there is no reason to undertake the infinitely complex project of comparing all possible worlds. If on the other hand, the proposition *does* make an existence claim (either explicitly or implicitly), -- then it is impossible to. determine the truth-value of that proposition[28] without knowing whether the possible world inhabited by the subject term is or is not the best. So this type of infinite analyzability requirement is a reliable *sign* of contingency.

It is not, however, the *source* of contingency. The length of the analysis is not what guarantees contingency, but rather the supposition that even if the analysis is completed, God is free to decide whether or not to create the best possible world. In other words, even if the analysis were completed and even if it were demonstrated that the subject in question is part of the best possible world, this would in no way entail that it exists. For being a part of the best possible world, while it is a necessary condition for existence, is not a sufficient condition. Another premise is required: that God created the best of all possible worlds. This premise is *not* contained in the analysis of the subject term, even if the subject term is completely unpacked. The infinite analysis reveals a *ground* for existence; it does not reveal existence itself. It is *not* the length of the analysis that makes a truth contingent but rather the fact that God was able to make a free decision on the basis of that analysis.

Leibniz maintained that there is an *a priori* ground for the truth of every proposition, contingent or necessary:[29]

> All contingent propositions have reasons for being as they are rather than otherwise, or (what is the same thing) they have *a priori* proofs of their truth, which render them certain, and show that the connection of subject and predicate in these propositions has its foundation in the nature of the one and the other; but they do not have demonstrations of necessity, since these reasons are only founded on the principle of contingency, or of the existence of things, i.e, on what is or appears the best among several equally possible things.

It is a mistake to move from the claim that analysis reveals a reason for the truth of every proposition to the claim that all propositions are analytic. Contingent propositions (those which assert existence) are not analytic for Leibniz in any usual sense of that word: (1) The subject does not contain the predicate (even if we construe existence as a predicate). (2) To deny a contingent truth involves no contradiction. And (3) contingent truths are not reducible to logical truths in any way.

If we are to maintain, therefore, that existential propositions are analytic for Leibniz, then we must mean something very different by "analytic" from what we usually mean. Surely the most charitable rational reconstruction of Leibniz's theory is one which maintains that although analysis always reveals a ground for the truth-value of a proposition, only non-existential propositions are truly analytic.

One last point remains to be made on this topic: The account which maintains that infinite analyzability is the root of contingency makes the distinction between necessary and contingent truths an *epistemological* rather than an ontological distinction. The difference between a finite and an infinite analysis, although crucial for finite minds, would not appear to be very significant for the divine mind. (It doesn't even make sense to say that a finite analysis is "easier" for God than an infinite analysis.)

But if this is true, then in the eyes of God, all existential propositions are in the final analysis (so to speak) of the form AB is A (where A stands for existence). If this were true, then the subject could not have failed to exist (for even God is bound by the laws of logic).

We may want to hedge our bets a bit by maintaining that no true proposition is, according to Leibniz, "radically synthetic" if by "radically synthetic" we refer to propositions in which a totally arbitrary predicate is attributed to a subject. Contingent propositions are, for Leibniz, only "moderately synthetic": something new is attributed to the subject, but there is within the subject itself some reason for this attribution.

If we insist upon foisting the analytic/synthetic distinction onto Leibniz's theory, then we must say that it cuts the same way for him as the necessary/contingent distinction. In fact, contingency is rooted in the very circumstances that the truth-values of some propositions are *not* analytically determined and that God, therefore, does have some latitude for providential decision-making. It is this device which safeguards the absolute contingency of laws of nature. We turn now to the question of their hypothetical necessity.

The Hypothetical Necessity of the Laws of Nature

We have already established that because relationships among essences are matters of logic; even God cannot tamper with these (according to Leibniz). Hence, we can say that once God decided to actualize a maximum of entity, the creation of this world followed inexorably. So, of course, did the *laws* of this world. If we let *p* represent God's decision to create the best possible world and if we allow *q* to represent the laws which obtain in this world, we can say: Necessarily, if *p* then *q*. ($\Box(p \supset q)$) What is *most* important to realize at this stage is that we are not entitled to say: If *p*, then necessarily *q*. ($p \supset \Box q$)

This will be more comprehensible if we consider a specific law obtaining in this world which says that event-type *r* is always followed by event-type *s*. ($r \supset s$) We can legitimately say $\Box(p \supset (r \supset s))$. In words, this states that necessarily, if God decided to create the best possible world, then this law obtains. What we are *not* entitled to assert is the following: $p \supset \Box(r \supset s)$. This says that if God decided to create the best possible world, then there is some *necessary connection* between events of class *r* and events of class *s*. This is much like the geometrical necessity of Cartesian mechanics according to which laws of physics embody necessary relationships -- i.e., relationships among essences.

When Leibniz denies the legitimacy of the last assertion, he is denying that there must be some necessary relationship among events linked by physical laws; ***there need be no strict entailment among them.*** The arrow of causation is not the arrow of entailment. Though it may be absolutely certain that *r*-events will always be followed by *s*-events, it need not be the case that the two classes of events are ***essentially*** related. In another world, events of class ***r*** might ***not*** be followed by events of class ***s***. Laws of nature, though hypothetically necessary, embody only ***contingent*** connections.

This is an extremely important distinction. It allows Leibniz to say that even now (long after the creative decision), the future is in a very significant sense contingent, though perfectly determined.[30]

> The whole universe might have been made differently; time, space, and matter being absolutely indifferent to motions and figures; and God has chosen among an infinity of possibles what he judged to be the most suitable. But as soon as he has chosen, it must be admitted that everything is comprised in his choice, and that nothing can be changed, since he foresaw and arranged everything once for all....It is this necessity which can be attributed now to things in the future, which is called hypothetical or consequential....But though all the facts of the universe are now certain in relation to God,...it does not follow that their connection is always truly necessary, i.e., that the truth, which pronounces that one fact follows from another, is necessary.

Clearly, this distinction allows Leibniz to distance himself even further from what he regards as the necessitarianism of Cartesian mechanics. For Leibniz, as much as for Kant, it is true that "the wind does not follow the rain on account of the law of contradiction." The concept of the effect is ***not*** necessarily contained in the concept of the cause.

The Possibility of *A Priori* Knowledge of Physical Laws

If then it is acknowledged that the laws of nature do not embody ***conceptual*** linkages for Leibniz, then how -- it may be asked -- can he possibly maintain that the laws of nature are sometimes susceptible of ***a*** *priori* discovery?

It will be argued here that just as it is a mistake to assume that the Leibnizian notion of anlaysis can be immediately translated into our notion of "analytic truth", so also it is a mistake to assume that the Leibnizian meaning of "*a priori*" is equivalent to ours.

A word first about the role of scientific laws or principles of order in God's creative act: We have seen that God's aim in creating the world was to maximize being. This is fine until we notice that Leibniz sometimes seems to suggest that the creator has a second goal as well; to create that world which is governed by the simplest laws. It is very easy to imagine situations in which these two aims would operate at cross purposes. Indeed this would seem to be the rule rather than the exception. A world comprising a single undifferentiated particle would no doubt be governed by very simple laws while the infinitely varied and complex world which God chose to create would seem to require a much more complicated set of laws. It is easy then to form a picture of God playing off his two ends one against the other: giving up a little variety here for some symmetry there, swapping some particular essence for more overall simplicity.

This, however, is the wrong picture. The God of Leibniz did not have two conflicting ends. Rather the possible world governed by the simplest laws *is* the possible world which maximizes entity. This may be somewhat easier to understand if we consider the following analogy: Suppose I have a given amount of fencing material with which I hope to enclose a maximum amount of area. I can achieve this end by employing the "simplest" geometrical formula for a closed curve -- the formula for a circle. The "simplest" law in this case is also the one which maximizes area. Now this is all rather problematic, of course; but it does seem to capture Leibniz's idea:[31]

> The case is like that of certain games in which all the spaces on a board are to be filled according to definite rules, but unless we use a certain device, we find ourselves at the end blocked from the difficult spaces and compelled to leave more spaces vacant than we needed or wished to.

In another place, Leibniz argues that if God had chosen any but the simplest laws it would have been like constructing "a building of round stones which leave more space unoccupied than that which they fill."[32] So selecting the world with the simplest laws of nature is a way of selecting the best possible world.

When Leibniz says that in principle all propositions admit of an *a priori* proof, he means that all propositions can be explained either with reference to the truths of logic or with reference to God's plan for the world. And when he says that sometimes we can discover an *a priori* proof for a law of nature, he means that we may sometimes be led to discover a law of nature by considering God's plan for the world.

The basic picture at work here is that of a three-layer explanatory structure, the middle layer of which comprises scientific laws. The top layer tells us that God decided to maximize perfection when he created the world and that he did so by selecting the simplest possible laws. The bottom layer comprises all the physical phenomena which we as human scientists wish to explain. So the top layer of the structure, the principle of perfection, explains why the laws of nature obtain; and the laws of nature in turn explain the order of phenomena. The first sort of explanation is teleological; it is explanation in terms of God's purpose. The second sort of explanation is in terms of efficient causation.

As scientists, our job is to discover the laws of nature. Our epistemological situation as human knowers is such that we must ascend (so to speak) from below. We discover laws of nature by carefully observing the phenomena, keeping detailed records, paying special heed to samenesses and differences, and so on. We can employ this *a posteriori* methodology with complete confidence, knowing that every phenomenon admits of an explanation in terms of efficient causes (even animal motions).

There is, however, an alternative methodology which is at least potentially available to the very best natural philosophers. This methodology would allow us, at least in principle, to reason *down* from God's decision to create the best possible world to a statement of the laws he selected. It is this methodology which Leibniz terms *a priori*. Notice that it is *not* incorrigible and it is *not* even entirely independent of sensory experience. Leibniz emphasizes clearly the possibility of error when we attempt to employ this method:[33]

> I admit, however, that though this way is not hopeless, it is certainly difficult and that not everyone should undertake it. Besides, it is perhaps too long to be covered by men. For sensible effects are too greatly compounded to be readily reduced to their first causes. Yet superior geniuses should enter

> upon this way even without the hope of arriving at particulars by means of it, in order that we may have true concepts of the universe, the greatness of God, and the nature of the soul....Yet we believe that the absolute use of this method is conserved for a better life.

In this life, its use is hypothetical. The *a priori* methodology of final causes functions largely in a heuristic role -- it allows us to propose conjectures about the general structure of scientific laws, and it sometimes gives useful direction to our research. It was mentioned earlier that *a priori* methodology of final causes is not entirely independent of experience. This is because one of the ways we come to understand the nature of God and his purposes is by doing science. So the more we have learned experientially about nature, the more adept we should expect to be at employing the *a priori* method. For example, if we know that God sometimes employs least-time laws in ordering optical phenomena, we will be alert for opportunities to employ variational principles in other domains as well.

On the subject of variational principles, it should be noted that even in the twentieth century some prominent physicists have thought that variational principles give evidence of design in nature. Even though the differential and integral forms of a law are mathematically equivalent, the two forms of expression seem to give rise to different ways of speaking. Because the differential form of a law expresses a subsequent state in terms of an immediately preceding state, it is often discussed in terms of efficient causation. The integral form, on the other hand, gives each particular state in terms of the total process, thus giving rise to the appearance of teleology and to discussion in terms of final causes. This distinction is very similar to Leibniz's distinction between *a priori* and *a posteriori* methods for explaining the same law.

The hypothetical and experiential aspects of the *a priori* methodology have been emphasized here in order to counteract the standard textbook picture of Leibniz as an armchair scientist. Despite the efforts of Gerd Buchdahl[34] and others to show just how inadequate this picture is, it continues to be the dominant one.

Conclusion

The exposition is now complete. It has been shown how, why, and in just what sense Leibniz believed that the laws of nature are (1) absolutely contingent, (2) hypothetically necessary, and (3) *a priori* deducible, at least in principle. It has also been argued that it is incorrect to maintain that the laws of nature are analytic, even though it is true to say that an infinite analysis would reveal reasons for their obtaining. Leibniz believed that these complex modal and epistemological distinctions enabled him to negotiate a safe passage between the necessitarianism of Descartes and Spinoza and the arbitrariness of Newton.

The most striking feature of all these distinctions is that they are grounded in both theology and dynamics. The theological aspects have been emphasized here, but it would be foolish to attempt to separate the two.

There is not room here for any bogus distinction between internal (i.e., scientific) and external (non-scientific) factors influencing Leibniz. The two are inseparable. As he himself said: "I begin as a philosopher, but I end as a theologian."[35] It would be equally correct to say that when he begins as a theologian, he ends as a natural philosopher.

Department of Philosophy
University of Western Ontario

Notes

Abbreviations

L	*Gottfried Wilhelm Leibniz: Philosophical Papers and Letters*, ed. by L. Loemker (Dordrecht, 1969).
G	*Die philosophischen Schriften von Leibniz*, ed. by C.I. Gerhardt, 7 vols. (Berlin, 1875-90)
C	*Opuscles et fragments inedits de Leibniz*, ed. by L. Couturat (Paris, 1903)
Grua	*G.W. Leibniz: Textes inedits*, ed. by G. Grua (Paris, 1948)

1. Spinoza, *Ethics*, Part 1, proposition 33, translated by R.H.M. Elwes, (New York: Dover Publications, 1951).
2. L, p. 680. (The controversy between Leibniz and Clarke, Clarke's second reply, second paragraph.)
3. L, p. 687. (Leibniz's fourth letter, second paragraph.)
4. Margaret Wilson, "Leibniz's Dynamics and Contingency in Nature" in Machamer and Turnbull (eds.) *Motion and Time, Space and Matter* (Ohio State University Press, 1976).
5. *Ouvres de Descartes*, ed. by C. Adam and P. Tannery (Paris: Leopold Cerf, 1897-1910), 6; 43.
6. In Harry Frankfurt (ed.) *Leibniz: A Collection of Critical Essays* (Garden City, N.Y.: Doubleday, 1972) pp. 69-97.
7. G. v. 286.
8. C. 522.
9. Bertrand Russell, *A Critical Exposition of the Philosophy of Leibniz* (London: Allen & Unwin, 1900: 2nd ed. 1937) p. 67.
10. C. 9.
11. G. vii. 303.
12. C. 376.
13. See p. 186 of 1903 review of Couturat: "Recent Work on the Philosophy of Leibniz", *Mind* xii (1903).
14. Louis Couturat *La Logique de Leibniz d'après des documents inédits* (Paris: Alcan, 1901) and "Sur la métaphysique de Leibniz", *Revue de Metaphysique et de Morale*, 10 (1902)

15. C. 9.
16. G. vii. 303.
17. *Leibniz: Textes inédits*, edited by G. Grua (paris, 1948) p. 304.
18. G. vii. 195 n.
19. C. 23.
20. G. vii. 309.
21. G. ii, 181.
22. G. ii. 423.
23. Grua, p. 302.
24. Wilson, *op. cit.*, p. 285.
25. G. vii. 309.
26. C. 20.
27. G. ii. 49n. For a discussion of this topic, see G.H.R. Parkinson, *Logic and Reality in Leibniz's Metaphysics* (Oxford: Clarendon Press, 1965).
28. Assuming, of course, that it does not embody a contradiction.
29. G. iv. 438.
30. G. ii. 400. See also G. vi. 123.
31. L, p. 487. ("On the Radical Origination of Things").
32. L, p. 211. (Letter to Malebranche, June 22/July 2, 1679).
33. L, p. 283. ("On the Elements of Natural Science, part II: An Introduction to the Value and Method of Natural Science").
34. G. Buchdahl, *Metaphysics and the Philosophy of Science*, Blackwell, 1968.
35. Eduard Bodemann, *Die Leibniz-Handschriften der Koniglichen offentlichen Bibliothek zu Hannover*, Hildesheim, Georg Olms, 1966, p. 58. Cited in Curley, *op. cit.*

Robert E. Butts

LEIBNIZ ON THE SIDE OF THE ANGELS*

The Methodological Angel

In the thought of Leibniz there are many roles that angels play; he also thought that angels play many roles. For example, he thought that angels are miracle-workers: they can perform "inferior" miracles; they can make a man walk on water without sinking [*4th letter to Clarke*; Loemker #44].[1] This is a role that an angel can play. It is not a role in the thinking of Leibniz; it is something an angel can do. I do not want to discuss things that angels are thought to be able to do, but rather things that angels do in the thinking of Leibniz: roles they may be thought to play in Leibniz' reasoning about things, hypothesizing about things, concluding or doubting about things. We may call angels that Leibniz thought could do various things *ontic* angels, and angels in the thinking of Leibniz we will call *methodological* angels. As we will see methodological angels do not *do* anything; they function as possibilities, as voices we might hear. It is true than an angel can say something to us based only on what it is; the two kinds of angels are connected: possibility requires actuality. But just now I want to discuss a major role Leibniz thought that methodological angels, not ontic ones, can play.

---

*This is a revision of the conference paper. The material is put to full use in Chapter II of my *Kant and the Double Government Methodology, Supersensibility and Method in Kant's Philosophy of Science*. Dordrecht, Holland: Reidel, (1984)

K. Okruhlik and J. R. Brown (eds.), The Natural Philosophy of Leibniz, 207–226.

Angelic Explanation

In an early fragment[2] concerning the proper method of investigating the causes of natural things (things pertaining to bodies), Leibniz starts the discussion with this very curious axiom that he takes to be certain:

> *...all things come about through certain intelligible causes, or causes which we could perceive if some angel wished to reveal them to us.*

What is it exactly that an angel could reveal to us? What can we not perceive that an angel can? We know from other texts of Leibniz that perceptions are *states of perspective*, "points of view". They are oriented states of consciousness which are either clear or confused, accurate or distorted. We "move" from one such state to another through *appetition*: our orientations, however confused or distorted, are directed. This means, among other things, that we change states of orientation because we have needs, wants, expectations. Let us suppose that our lack, in the case in hand, is a lack of understanding. For example, we do not understand why iron filings gather around a magnet in the way they do. Leibniz' axiom tells us that this phenomenon has an intelligible cause knowledge of which would give us understanding. We could understand that cause if an angel were willing to reveal it to us. What is it that the angel would reveal?

I think it would reveal its own state of much greater clarity -- its own *perfect* orientation with respect to cases of magnetic attraction. God perceives perfectly; his perspective is flawless.[3] God's *logical situation (situs)* is perfectly centered.[4] Angels come close to this perfect perception in the case of some ideas. If we could share angelic clarity in some of our ideas, we would know -- accurately perceive -- true intelligible causes.

In the next sentence of the fragment Leibniz tells us that since we can "accurately perceive" only "magnitude, figure, motion, and perception itself," everything must be explained through magnitude, figure, motion and perception. In this fragment Leibniz is concerned with those things that are without perception, what we would call *physical events* (his examples are reactions of liquids and precipitations of salts). Lacking perception -- being material object transactions rather than orientations of consciousness -- these things can only be explained by reference to magnitude, figure and motion.

Again: what would the angel be helping us to understand if he were to undertake to "explain" these cases to us? It would seem that, along with the Cartesians, Leibniz is accepting that we can have "clear and distinct" ideas of magnitude, figure and motion, that we can clearly perceive them.[5] The methodological angel would, I think, urge upon us *the doing of science.* He would insist that if we are to understand these cases of physical action, we need *to reduce* the phenomena involved to quantitative aspects of the states of the objects. And those properties of the states of reaction, precipitation, declination that are quantitative are precisely the properties of magnitude, figure and motion.

Galileo and Plato

It seems inescapable that we conclude that Leibniz' angel accepts a methodological form of the distinction between primary and secondary qualities. The angel is trying to persuade us to accept that we must explain all events which are without orientation in mechanical terms. What cannot be so explained must "be referred to the action of some perceiving being." These properties of the material object transactions that cannot be reduced to mechanical causes -- cannot be accurately perceived or oriented in perspective -- are precisely the secondary qualities so-called, qualities like the taste of the object, or its colour. It is important to note that such qualities are not reducible to mechanical ones -- are not expressible in numbers denoting magnitude, figure or motion. What the angel must do for us, then, is help us to find ways *to discount* these very badly distorted perceptions of the material objects.[6]

The methodological angel "must explain some cause to me, such that, if I understand it [achieve accurate perspective], I can see that the phenomena follow from it as necessarily as the cause of the hammer stroke when a given time has elapsed follows from my knowledge of the clock." Thinking as the angel invites me to think I must find ways *to control* features of the material objects, the bodies, so that the information I attend to yields real causal knowledge. I must, as Galileo put it, "discount the material hindrances," subtract the inessential, focus only upon what is genuinely in perspective, only upon what can be seen as the angels see it.[7]

When applied to our problems in attempting to understand the behaviour of bodies, angelic thinking becomes unremittingly *experimental.* The experiments, so Leibniz says, "*are to be carried out isolatedly,*" so that "nothing enters into the process but the general and necessary agents." To think as an angel thinks -- or would have to think if he had our kind of body -- is to reduce body to number by means of a methodology Leibniz calls "mechanical"; it is to analyze compounds into simples according to a rational scheme implementation of which includes perfecting the instruments employed in the investigation. For Leibniz, method and technology go hand in hand. An angel thinking in numbers and looking through the lens of a microscope while reading the face of a clock -- that, for Leibniz, is a methodological angel.

The God's-eye View

I propose to delay further discussion of the mechanical method the angels wish us to adopt, and to relate what we have discussed so far to other well-known doctrines of Leibniz.

We have seen that what the angel would tell us is that we are to study the behaviour of bodies through controlled experiments directed toward obtaining knowledge of necessary causes. To adopt this methodology is to achieve more perfect orientation or perspective with respect to bodies. The general methodology of the angels is related to the following views of Leibniz: (1) the centrality of the substance/accident category; substances are real; bodies are only phenomenal, they are what Leibniz elsewhere calls "semibeings".[8] (2) All propositions asserted in relational form are to be reduced to propositions about substances -- or all relations are to be reduced to substances. (3) Although explanations referring to magnitude are (as we have seen) in some sense fundamental, even these explanations are in principle further reducible to explanations in terms of formal or qualitative features Leibniz calls "situations".[9] He promised, but did not deliver, a formal analysis of the geometry of situation, and I do not think it is often enough realized that this geometry would in effect be the geometry of God. It is a geometry of *perspectives*, and such points of view (being states of substances) are more basic than magnitudes, figures or motions. They are states of perception.

Leibniz himself provides the apt metaphor for capturing the relations of his proposed methodology to other basic doctrines. It is the metaphor of the plan of a city.[10] More specifically, it is the plan of a city looked down upon from the top of a high tower placed in its centre. He says that this plan differs from the ones we would get "from the almost infinite horizontal perspectives" obtained from entering the city from one direction or another. He also says that "This analogy has always seemed excellently fitted for understanding the distinction between nature and accidents." Seen from the high tower the perception of the city is perfect -- centered in a true substance. Seen *from any other* point the perspective on the city is distorted -- presenting an accidental, semireal plan.[11]

Leibniz develops the metaphor in even more apposite ways. He says that to see the city from the high tower is "as if you intuit the essence itself." Approaching the city in other ways (from without) is "as if you perceive the qualities of a body." He presses on: "And just as the external aspect of the city varies as you approach it differently, from the west or from the east, the qualities of a body vary with the variety of our sense organs." Our angel sees the city, if not from the exact top of the high tower, certainly from a vantage point close to the top. What he tells us is to so orient ourselves with respect to the city that we approximate the view from the tower as closely as is possible for us, and this means that we try to give explanations of the behaviour of bodies that are explanations based on clear ideas of magnitude, figure and motion, rather than on information obtained directly from sensation.

Thus understood, mechanical forms of explanation are attempts to reduce relations between semibeings (bodies) to the substance/accident category. It is this form of explanation that provides intuition of the essential features of bodies (the primary qualities). The "almost infinite horizontal perspectives" one can take on the city correspond to the variety of our less well oriented perceptions arising through "the number of our sense organs." Each sense organ can be viewed as a potential generator of very many possible perspectives along various scales of degrees of intensity, duration, "size", and other parameters of sense perception. Each determinate instance of a parameter expresses the "feel" of a possible definite sensation. The totality of all possible sense organ-generated perspectives is nearly infinitely large, a fact that Leibniz interprets as meaning that the likelihood that sensations will ever yield true perspectives (true stories about reality) is vanishingly small.

Empirical Adequacy

Each substance contains a potentially infinite number of accidents (in the one true substance -- God -- the infinity is actualized); furthermore, it contains them necessarily: each substance is a mirror of *all of* reality; each substance, from its own point of view, lives its life history as a form of the life history of every event that could possibly happen. By analogy, there are an infinite number of possible plans of the city, or an infinite number of possible sense data accounts or descriptions of the world. On Leinbiz' view it would not be possible *to choose* between these various accounts in any *empirically adequate* way; the method of science must transcend the empirical particular; a scientific account must selectively orient our perspective on things by reduction of secondary to primary qualities, by reduction of relations to substances -- in general, by Platonic reduction of specific content to form.

As we might expect, Leibniz approaches the problem of empirical adequacy of theories at different, but related, levels. The mechanical method -- to which we will return presently -- does deal with the problem of adequate explanation in science. Elsewhere, Leibniz addresses the problem as one of distinguishing between the real and the merely imaginary. The setting of this problem is quite Humean (or Cartesian -- is there a difference?): I am a mind with a variety of phenomenal contents -- how to distinguish between those that indicate real things and those that are only dreamt up?[12]

Leibniz thought that I cannot demonstrate criteria of the empirical adequacy of purported real explanations of phenomenal behaviour; I can achieve moral, but not metaphysical, certainty. "...By no argument can it be demonstrated absolutely that bodies exist, nor is there anything to prevent certain well-ordered dreams from being the objects of our mind, which we judge to be true and which, because of their accord with each other, are equivalent to truth so far as practice is concerned." The criteria of empirical reality that produce moral certainty or "greatest probability" are these:

A given phenomenon must be:

1. *vivid* in intensity of sense quality;
2. *complex* in variety of its qualities, leading us to undertake many experiments and make new observations;
3. *coherent*, in that it is made up of many phenomena for which reasons can be given, *and* it conforms to the regularly and repeatedly observed orders and positions of similar phenomena;
4. *coherent*, if, together with other phenomena, it can be referred to a common cause;
5. *in consensus with* the "whole sequence of life," shared with

and communicated to others;

6. *predictable*: "Yet the most powerful criterion of the reality of phenomena, sufficient even by itself, is success in predicting future phenomena from past and present ones, whether that prediction is based upon a reason, upon a hypothesis that was previously successful, or upon the customary consistency of things as observed previously."

If we were to strip away Leibniz' commitment to the ultimate primacy of metaphysical explanation, would he turn out to be a van Fraassen anti-realist or a Hessean instrumentalist? I think perhaps he would. Fortunately for Leibniz his angel in thought will prevent us from performing the operation. Leibniz' six criteria of empirical reality (or of the empirical adequacy of an explanation) seem to me to reduce to two: the internal richness, complexity and interest of the given phenomenon; and the external acceptability of the phenomenon as a coherent ingredient in the individual and shared *commonness* of things. Miracles, the *uncommonness* of things, are for God and the ontic angels. The realm of bodies is the realm of the familiar.[13]

Mechanical Methodism

We thus have Leibniz' restricted assurance that we need not go mad -- there is a kind of reality characteristic of bodies that we can take with us into court, into the casino, and, with the help of our angel, into the laboratory.[14] Under the protection of the law, with our pockets bulging with gold, we can go forth confidently to engage in science. We can investigate the world machine, a machine as regular as a clock, as interesting as a rainbow, as ordinary as a water fountain that will not flow. We have an angel to take us by the hand.

What I will be seeking in the laboratory, remember, is causal knowledge, knowledge of mechanical connections. I must know in the strongest possible sense that the hammer stroke must fall when a certain time elapses, a fully certain cognition that follows from my knowledge of the internal mechanism of the clock. The attainment of such knowledge of causal mechanisms presupposes quantified information about distances and times; it presupposes the possibility of locating objects in places occupied during elapsed or elapsing times. From other texts we have learned about Leibniz' relational theory of space and time. We know that for him space and time, like phenomenal objects (bodies), arise only from confused perception. Let me remind us in outline of some features of Leibniz' thoughts about space and time.

For Leibniz all ideas that are not clear and distinct (well oriented) arise through confused perception -- thus for our ideas of space and time: we see things incompletely, we see them, for example, beside one another, at distances from one another, coming one after another; in short, as extended in various forms and from time to time. Next, by a process of abstraction, we think of the places and moments occupied by the objects perceived. From the actual places and moments thus abstracted, we go on to abstract possible places and possible moments that any objects might from time to time occupy. These *formats* of possible places and moments (instants) are space and time.

Space is thus seen as a form of possible spatial relations and time as a form of possible temporal relations; as formal matrices space and time are *ideal*, not real. They are idealizations of places and moments, not actual things. Thus, for Leibniz, they cannot be substances, nor accidents of substances. They can therefore only be relations. This means that in order for there to "be" space and time, there must be things in space and time, since space is *only* the spatial relations between relata and time is similarly *only* the stretches of time between and involving events. No relata as objects in places, no space. Similar considerations hold for time and events.

These considerations yield Leibniz' relational theory of space and time, as follows:

1. Objects are independent of and ontologically prior to space; there is no concrete *entity* space. The same is true for time and events (or processes).
2. Absolute motion is therefore unintelligible; all motions of objects must be relativized to aspects of the motion/rest of other objects. And similarly for absolute time.
3. Space and time are only given as (sensible) measures. There are no "properties" of space and time. There are only spaced objects and timed events.

Notice that none of this has anything to do with the "real" space and time of the MONADS. The "space" of the monad is its logical place (its situation). The time of the monad is remembering and expecting. Ideal space and time "mirror" these monadic structures, but do not exactly "map" them. Idealized space and time are abstract formats (structures of possibility); real space is perspective (*situs*), and real time is directedness (*entelechy*).

In these brief remarks on space and time in the thought of Leibniz I have perhaps been reminiscing about the obvious. Leibniz' theory about space and time as relative is after all part of his metaphysics, and this theory, although perhaps presupposed by it, is not a direct concern of Leibniz when he discusses the mechanical method.

One matter of interpretation that I want to be on my guard about, however, has to do with confusion of space and time as ideal formats, and bodies as phenomena. Space and time and phenomena are for Leibniz semibeings, meaning that they are none of them substances or accidents of substances. But there is a difference between a realtion and a phenomenon, and that difference is important for Leibniz (as it will be later for Kant). A relation is a confused pattern of logic that must be "reduced" to substance; a phenomenon is a confused intentional object that requires clarification. It is the business of metaphysics to achieve the reduction of relations to substance, and it is the quite different business of the mechanical method to clarify our awareness of bodies.

[In public discussion of this paper J.E. McGuire argued that the common interpretation of Leibniz as a metaphysical "reductionist" is likely mistaken. I think he is correct if he means us to take Leibniz as having held that there is literally no sense in which a relation like space or time can be reformed as a substance. Strictly speaking semibeings like space and bodies are beings of an imperfect sort, but beings nevertheless. We can achieve a legitimate knowledge of the behaviour of bodies acting in space and time, a kind of knowledge *perfected by* the mechanical method and *justified by* the metaphysics of the monadology. But the rules of the method and the principles of the metaphysics do not provide ways in which space/time observation sentences can be translated into more basic substance/accident statements. The most I can expect from the rules and principles is some assurance that in the best cases my knowledge of the physical world is as perfectly attuned to that of the angels as is humanly possible. To make my case as persuasive as possible I would need to go into the details of Leibniz' ideas of the "well-founding" of phenomena, but that is not a detour which can be helpfully followed at this point.]

The mechanical method will yield "truths" given expression in the relational formats of space and time. But achievement of contingent truth is, strictly speaking, atemporal and without place. An experiment, after all, is a deliberate attempt to diminish the effects of time and place. To achieve angelic clarification of our ideas of objects behaving from place to place and from time to time involves seeing how close we can come to removing the objects from the contraints of being somewhere at some time.

Angelic Alchemy

We have learned that the angel urges upon us experimental investigation of nature. Experiments are to have the force of reducing behaviour of phenomena to information cast in terms of necessary causes and given expression in the epistemic syntax of quantity, figure and motion. The grammar of causal explanation is mathematics; the clear perspective on bodies attainable by human beings is quantitative in character. The best embodiment of such quantified causal information is of course physics. But mathematical physics of the sort Leibniz actually developed is already a reduced and clarified perspective on bodies -- the effects of uncontrolled sense experience are perfectly managed in physics, or very nearly so. The mechanical method best represented in physics presupposes a prior method as the basic method of administering and organizing sense data. In order to outline this more basic method Leibniz is led to discuss examples drawn largely from chemistry.

In his 1677 *On a Method of Arriving at a True Analysis of Bodies and the Causes of Natural Things*, Leibniz cites as examples chemical composition and preparation (sulfur arising from vitriol); knowledge obtained through empirical exercise of a craft (coal fire differs from the fire of a torch); and manufacture (ironmaking). These cases of phenomena drawn from the low sciences are introduced deliberately as involving basic "modifications", a partial list of which includes modifications of "weight, elasticity, light or heat, coldness, liquidity, firmness, tenacity, volatility, fixity, solubility, precipitation from menses, crystallization". I think that Leibniz' methodology for studying bodies is an extension of practices in the alchemical laboratory, with the method of analysis of "modifications" presupposing the theory of combinations in his *De arte combinatoria* of 1666.

How much distance is there between taking the results of alchemical experiment to be related to positions of heavenly bodies and in turn to events in human history as *signs* are related to the things signified; and taking the results of any kind of controlled experiment as purifications of contaminated sense experience, and hence as *evidence* for physical hypotheses?[15] Let it suffice in this place to suggest that in their way, alchemists were as distrustful of sense qualities (the secondary qualities) as were Galileo, Descartes and Leibniz in their ways. However, whereas it was probably typical of alchemists to confuse states of physical objects with sensible qualities -- for an alchemist is coldness a property of a body or a sensation? -- the great strength of Leibniz' suggested methodology is that the angel tells him clearly that qualities of bodies are different from sensible qualities.

The list of modifications quoted above is a list of essentially *dispositional properties*, they are states in which from time to time we might find bodies. As possible states of a physical object (a body) they are expressible as numbers, as measurable features of that object in a certain state. For Leibniz there are two kinds of analysis of bodies, one, an experimental determination of various physical qualities; the other, the ratiocinative resolution of sensible qualities into causes or reasons. This is important, because Leibniz is suggesting a methodology that is more complex than it might at first sight appear.

He says, "So when undertaking the most accurate reasoning, we must seek the formal and universal causes of qualities which are common to all hypotheses and must begin accurate but universal enumerations of all possible modifications..."[16] The first stage in proper empirical method is thus the rational development of possibilities, taking due account of what we are given in sensation. Any state that we might find a body "really" to be in is first suggested to us through philosophizing about *the kinds of states objects must be in if we are to account for our having the kinds of sensations we in fact have.* Each of Leibniz' "modifications" turns out to be a measurable state of an object correlated with some actual sensible condition I am in at some time. In the best scientific cases the sensible condition will be "reduced" -- effectively eliminated -- through appeal to experiemntal data that are cleansed of the idiosyncracies of sensation.

Thus I sense heat and cold, "feel" firmness and elasticity. Each such state of sensation, *properly thought about,* provides awarness of a possible "real" state of a body whose causes become accessible to me through experiment. [Throughout all of this, of course, the angel is whispering in my ear.] The experiment, if you will, *improves* my sensation, just as a microscope improves my perception of the "invisible" components of an object.[17] Indeed, Leibniz also thinks that kinds of sensation can themselves be efficacious in helping us to approach better knowledge of causes. Thus he believes that "there is no medium more effective than taste for discerning the essential nature of bodies, because taste brings bodies to us in their substance and dissolves them in us so that we may perceive the whole solution closely".[18] [This suggests that in the laboratory of Leibniz there will be many dead experimenters! Surely he would have welcomed introduction of a *tasting instrument* incapable of being poisoned.]

Angelic Logic

It begins to look as if for Leibniz the ascertainment of real causes results from extension of the mechanisms of sensation. And this is as it must be: bodies are given in sensation as distorted perceptions or perspectives. What the angel gives us is microscopic vision -- but it is still *vision*. We are now in a better position to appreciate what it is that the angel tells us. We can now understand just what is involved when "he...explain[s] some cause to me, such that, if I understand it, I can see that the phenomena flow from it as necessarily as the cause of the hammer stroke when a given time has elapsed follows from my knowledge of the clock."

The clock is a machine constructed on certain principles and programmed to activate hammer strokes at measured intervals. The occurrences of the hammer strokes are exactly predictable if I know the plan of construction of the clock, the system of counting stretches of time, and the mechanical connection between the two. What causes the bell-ringer statues in the Piazza san Marco to "behave" as they do? Given license to do so I could perform experiments with the statues, taking apart the system in which they are parts. I could learn about the constituents of the system in this way, and something about the arrangement of the parts. But experiment unaided by the addition of geometrical principles, unaided by the addition of principles of logic, could not by itself yield knowledge of the cause of the action of the statues at regular times.[19]

There are at least two reasons why the angel would encourage us to supplement experiment with logic. First, a knowledge of arrangement and composition of parts of a body does not tell us why the structure is related to the gross behaviour of the body in just the way that it is. Second, as we have seen in the quoted text in footnote nineteen, for Leibniz all explanations are to be of the same type, causal principles must apply equally to both observable and unobservable phenomena. The causal logic of the ball leaving the hand of the player, or the statues in Piazza san Marco hammering the bells, is the same causal logic for clusters of insensible corpuscles colliding with one another. And it is only the principles of deductive logic that can lead me from cases of gross observable behaviour to cases of behaviour of unobservable entities.

If my state of knowledge could exactly match that of the angels, I could, Leibniz thinks, discover the interior constitution of bodies a priori from a contemplation of God. That is, from a knowledge of God as the author of all things -- from a knowledge of how God arranges his universe -- I could by deduction alone arrive at a perfect awareness of all causes of particular objects and events. Some progress along these lines is made by metaphysicians, but Leibniz concludes, "Yet we believe that the absolute use of this method is conserved for a better life."[20]

So far, Leibniz' angel has encouraged us to think that we can have accurate orientation with respect to manufactured devices like clocks, cases of contact action in observable situations (the ball and the hand), and collisions of unobservables like sub-microscopic dust. In all three cases the causal principles will be mechanical -- expressed in terms of magnitude, figure and motion. The angel had also promised that he would lead us to see how the secondary qualities can be "reduced" to mechanical principles. This is how Leibniz thinks the reduction works:

> Suppose that some angel wishes to explain the nature of color to me distinctly. He will accomplish nothing by chattering about forms and faculties. But if he shows that a certain rectilinear pressure is exerted at every sensible point and is propagated in a circuit through certain regular permeable or diaphanous bodies, and then teaches me exactly the cause and the mode of this pressure, and deduces the laws of reflection and refraction from it, thus explaining everything in such a way that it is clear that it could not even happen otherwise, then at last he will have increased my knowledge, since he has treated physics mathematically.[21]

We can now see the centrality of the case of the clock. The model of all proper physical explanation is the artifact, the manufactured object. All explanation of natural objects is to conform to this model. Thus our other cases -- ball/hand, sub-microscopic dust collision, and color -- are all to be cast in terms of the model of the planned and handmade object. The causal links in each case are all of them mechanical -- the preferred model is Cartesian: bodily behaviour is all a matter of contact action, impact. The "intelligible causes" that the angel would reveal to us if we listened are all mechanical, mathematically expressible, causes.

A Metaphysical Problem

We have come full circle in our quest for the wise words of the angel. We now know what it means to accept the axiom: "...I take it to be certain that *all things come about through certain intelligible causes, or causes which we could perceive if some angel wished to reveal them to us.*" It means that there is a method we humans can employ to clear our mental cavities, a method that will achieve for us an explanatory perspective close to that of the angels. There is much in the philosophy of Leibniz that is not clear, much that is equivocal. It is clear and unequivocal in his philosophy that the preferred methodology for studying bodies is the mechanical method. The problem is this: bodies are not real, only spirits are. Why is it that the clearest perception of bodies is not gotten by simply ignoring bodies altogether? In our four cases: machines, macroscopic contact action, sub-microscopic contact action, secondary qualities -- we get as close to the angels as we can possibly get by clear and distinct ideas employed in mechanical explanation. But perception is the real state of things, not mechanical action. From the aspect of eternity, purpose replaces efficient cause, reason replaces cause.

Why not go directly to states of teleologically organized spirit, following eagerly in the footsteps of Mary Baker Eddy? Why must we account for body at all? The answer, paradoxical as it may seem in the thought of Leibniz, is simply: *because there are bodies*. Bodies; not quite real entities, entities approaching the limit of the real. Finally, the requirement that there be bodies is, for Leibniz, a theological -- even partly mystical -- commitment. If I was correct in claiming that Leinbiz' metaphysics is at root a form of gnostic emanationism,[22] then it seems also correct to conclude that bodies are not quite unreal; they just approach the limit of the real: they are infinitesimally real.

It also seems to me to follow from Leibniz's principle of continuity that bodies -- as those beings approaching the limit of the real -- must exist and be accounted for. Remember, Leibniz thought that even angels have bodies, bodies not quite like yours and mine, but bodies neverthelesss. And if angels are in part distinct from God in having bodies -- which means, please bear in mind, *only* that angels more closely approach the limit of reality than does God -- then we, who are not yet angels, must surely have bodies.

Leibniz has assured us that in the "better life" [to come?] we will be able to know everything a priori as following logically from the nature of God. In this life we cannot achieve this perfect understanding; but we can approach and approximate it, and we can obtain some direct metaphysical assurances that in seeking to follow the advice of the angel, we are on the right path. We must account for the behaviour of bodies. That we can give an account of anything that is is axiomatic for Leibniz. We will not account for the behaviour of bodies if we contaminate that which is not spirit with things that pertain only to spirit. Leibniz' light of nature (his methodological angel) told him clearly that there are two orders, the mechanical and the formal (or spiritual); and that what is distinct in nature requires distinctly different explanation. The two orders of reality require two different forms of explanation.[23]

Not, of course, irresolvably different forms of explanation. We do approximate perfect perspective when we imitate the thought of the angels. But we cannot make a machine live, or an aggregate an organization, or a semibeing a being, or a phenomenon a real entity. In many places Leibniz warns us not to burden explanation of bodies by the introduction of substantial forms, operative spirits, or other kinds of occult, nonmechanical causes.[24] I believe this advice gives us good 17th-century scientific counsel to the effect that in pursuing our study of nature, we are not to expect to find occult causes, *hidden* forces.[25] For all their differences, Leibniz shares with Descartes and Newton -- and of course with his hero Galileo -- the dread of ascribing any form of scientific ignorance to God. Nothing is hidden from God, and not very much is hidden from the angels. Much is hidden from us, but only in reality, not in principle. There are forces that are not mechanical forces, but there are no irrational forces. Everything that is is transparent to perfect reason. Man and the angels are constrained to follow the "blind necessity" of mathematics -- blind, or maybe only blurred and astigmatic. Mathematical physics provides the prescribed corrective lenses; the mechanical method grinds those lenses.

A Speculative Postscript

The demon of Socrates, Descartes' light of nature, and the methodological angel of Leibniz: three powerful literary devices, or three profound philosophical masquerades? The demon of Socrates guarded him from moral error; the demonic messages were proscriptions. Descartes' impersonal light guided his reasoning by illuminating his intuitions. Leibniz' methodological angel stood ready to reveal necessary causes to him if he chose to listen, the revelations taking the form of rules to follow, advice to be heeded. The three masks hide wise counsel which would light our way to morally and epistemologically acceptable behaviour. Socrates faced a defamed popular morality; Descartes, a power-crazed theology and metaphysics; Leibniz, an irreverent and profaned science. For each, the root problem is soteriological. The sophists bought and sold morality; the 17th century Church enforced salvation; the alchemists demoted a sacred science to craft practices in the laboratory. Three public faces of profanity; three masks of salvation. For Leibniz, the task is finally one of returning science to the angels so that mere humans may be saved from ignorance. The dialectic is worthy of Socrates; the intuition, of Descartes. Leibniz promised to reinstate the sacred status of science, but of the "new" science, the science of machines and things of the earth. His alchemy would be the obstetrics of the world of body, performed by physicians clothed in the luminous robes of the angels.

Department of Philosophy
University of Western Ontario

Notes

1. Among the many other things angels can do: they can intuit truths directly with a degree of evidence that renders those truths indubitable [*New Essays*, 220-22]. The numbered Loemker references are to L.E. Loemker, ed., *Gottfried Wilhelm Leibniz: Philosophical Papers and Letters* (Dordrecht 1969). References to Leibniz' *New Essays on Human Understanding,* are to the P. Remnant and J. Bennett translation (Cambridge U.P. 1981).
2. *On the Method of Arriving at a True Analysis of Bodies and the Causes of Natural Things*, 1677; Loemker #15.
3. Leibniz' provocative image is that of surveying a town from the top of a tower positioned in its centre. This is the point of perfect perspective. The town viewed from any other point (in any other situation) is viewed imperfectly -- from the point of view of accurately situated perspective. I will show below that Leibniz makes other uses of this analogy.
4. *Studies in a Geometry of Situation, 1679*, Loemker #27. I will return to these points about perfect perception below.
5. Leibniz objected to Descartes' definitions of clear and distinct ideas; he did not object to clear and distinct ideas.
6. Discounting the effects of sensation is I think the fundamental form of "reduction" for Leibniz. It is not a matter of transforming phenomenal statements into "more basic" ones, but rather of achieving clearer orientation. In what follows I will use the word 'reduction' in several contexts, intending always that the word be understood in the sense of methodological transformation or epistemological re-orientation.
7. I have discussed Galileo's point in detail in "Some Tactics in Galileo's Propaganda for the Mathematization of Scientific Experience," *New Perspectives on Galileo*, ed. R.E. Butts & J.C. Pitt (Dordrecht 1978). Leibniz and Galileo share the impulse to reject secondary qualities as data for science against an accepted Platonic background. In the 17th century Platonism operated for some scientists as an abiding metaphysical paradigm.
8. *Correspondence with Des Bosses, 1709-15*; Loemker #63.

9. See footnote 4 above. In many places Leibniz insists that scientific explanations must in the end give way to metaphysical ones. For example: "Concerning bodies I can demonstrate that not merely light, heat, color, and similar qualities are apparent but also motion, figure, and extension. And that if anything is real, it is solely the force of acting and suffering, and hence that the substance of a body consists in this (as if in matter and form). Those bodies, however, which have no substantial form, are merely phenomena or at least only aggregates of the true ones." *On the Method of Distinguishing Real from Imaginary Phenomena*; Loemker #39. The uncompleted analysis of situation was to be Leibniz' formal logic of substance orientation.
10. Leibniz introduces the metaphor twice: *Letter to Jacob Thomasius, April 20/30, 1669*; Loemker #3; & *An Example of Demonstrations about the nature of Corporeal Things, drawn from Phenomena, late 1671*; Loemker #8.
11. The analogy obliquely expresses Leibniz' doctrine of reality as continuous, and his idea that creation is emanation from a single central source. However, the analogy is best fitted to capture some features of Leibniz' theory of knowledge and his theory of method. The related metaphysical concerns are discussed in Chapter I of my *Kant and the Double Government Methodology.*
12. My text is Leibniz' *On the Method of Distinguishing Real from Imaginary Phenomena* (date unknown); Loemker #39. Nicholas Rescher's *Leibniz's Metaphysics of Nature* (Dordrecht 1981), contains an interesting brief discussion of the fragment (pp. 13-15).
13. I am here presupposing what Leibniz thinks he can prove (Loemker #39), namely that there are minds other than my own, and that the total set of minds is appropriately interconnected. If the publicity of scientific results cannot be assured, the programme of the mechanical method surely fails, and the suggested nest of tests of empirical adequacy lies empty.
14. I am not convinced that we should also take it with us into the psychiatrist's consulting room! There remain places in which a sense of unreality had better prevail. Leibniz also knew about

those places.

15. This theme, although not discussed by him, was suggested to me while reading Ian Hacking, *The Emergence of Probability* (Cambridge 1978).
16. Loemker #15, p. 176.
17. In the fragment I am discussing Leibniz encourages us to employ the best technology available in our experiments: "*instruments of experimentation* -- scales, thermometers, hygrometers, pneumatic pumps -- and also ... vision, whether naked or fortified, ... smell, and, ... taste." This reference to technology, and others in Leibniz, thus appears to be fully justified in the larger context of his preferred methodology.
18. Loemker #15, p. 175.
19. Leibniz says: "*With the experiments are to be combined accurate and thoroughly extended reasonings after the manner of geometry, for only in this way can causes be discovered ...* Unless principles are advanced from geometry and mechanics which can be applied with equal ease to sensible and insensible things alike, nature in its subtlety will escape us. And reason must supply this important lack in experiment. For a corpuscle hundreds of thousands of times smaller than any bit of dust which flies through the air, together with other corpuscles of the same subtlety, can be dealt with by reason as easily as can a ball by the hand of a player." *On the Elements of Natural Science, Ca. 1682-84*; Loemker #32, pp. 282-83.
20. *On the Elements of Natural Science*, p. 283.
21. *Letter to Herman Conring, March 19, 1678*; Loemker #18, p. 189.
22. See my book cited in footnote 11 above.
23. The best statement of this double methodology is in *Critical Thoughts on the General part of the Principles of Descartes, 1692*; Loemker #42, pp. 409-10.
24. For example: *Critical Thoughts on the General Part of the Principles of Descartes, 1692*; Loemker #42, pp. 408-10; *Specimen Dynamicum, 1695*; Loemker #46, pp. 441-42; *On Nature Itself, or on the Inherent Force and Actions of Created Things, Acta eruditorium, September, 1698*; Loemker

#53. I try to shed light on some of these concerns in the book cited in footnote 11 above.

25. Leibniz fought the good fight against those who would contaminate the true Double Government Methodology. For example, in the Preface to *New Essays* (p. 68) we find him inveighing against those who "...saved the appearances by fabricating faculties or occult qualities, just for the purpose, and fancying them to be like little demons or imps which can without ado perform whatever is wanted, as though pocket watches told the time by a certain horological faculty without needing wheels, or as though mills crushed grain by a fractive faculty without needing anything in the way of millstones." I am indebted to my colleague, Dr. Howard Duncan, for this apt reference, which he himself employs to good purpose in his work on Kant and gravitation.

Jürgen Mittelstrass

LEIBNIZ AND KANT ON MATHEMATICAL AND PHILOSOPHICAL KNOWLEDGE*

1. Analysis and synthesis

Kant's comments on Leibniz are often marginal in form, but always essential in substance. It is in these comments that Kant distances himself from the philosophical tradition and establishes a new orientation in philosophy in an important way. This is also true of the reference to Leibniz in the (pre-critical) *Prize Essay (An Inquiry into the Distinctness of the Fundamental Principles of Natural Theology and Morals,* 1764), Kant's answer to the problem of the application of mathematical proof to the field of metaphysics posed by the Berlin Royal Academy of Sciences. The reference is of *epistemological* significance with respect to the system of the sciences. Here, Kant makes in a pragmatic form the distinction between mathematical and philosophical knowledge, which when it is later presented in the *Critique of Pure Reason,* in a systematically more elaborated form, forms an essential part of the 'transcendental doctrine of method'. The opposing party is, as Kant makes clear, Leibniz with his

*I would like to thank Robert E. Butts (The University of Western Ontario, London, Canada) for reading and making stimulating comments on the original version of this paper while staying at the University of Konstanz as Visiting Professor in 1983. He also helped me with the English text. In section 3 I have used some material from my "Substance and Its Concept in Leibniz" (see footnote 65), and in section 2 some material from my "The Philosopher's Conception of *Mathesis Universalis* from Descartes to Leibniz" (*Annals of Science* 36, 1979, pp. 593-610) first read at a conference in Benmiller's Inn on Lake Huron in 1978 -- organized by Robert E. Butts, patron of Leibnizian and Kantian scholarship.

K. Okruhlik and J. R. Brown (eds.), The Natural Philosophy of Leibniz, 227–261.

identification of both kinds of knowledge. From a systematic point of view, different ideals of knowledge and their realization in different disciplines -- paradigmatically given in conceptions of Leibniz and Kant -- are at stake.

Kant starts his attempt to answer the question how metaphysics is possible in the *Prize Essay* by stating that metaphysics is not possible as mathematics. At first sight, if one calls to mind the history of mathematics and philosophy, this looks trivial. Only the 'substantial' side, not the 'formal' side of this statement, is trivial. What is at issue is an essential difference in *concept formation* and, therefore, an essential difference with reference to the *constitution of objects* in mathematics and metaphysics or philosophy. Mathematics, as Kant says, starts to constitute its objects by giving their definitions ("in mathematics the definitions are the first thought"[1]); metaphysics starts with given objects or given concepts. The definitions of objects or concepts come not at the beginning but at the end of an imperative attempt at clarity which at the same time defines philosophy as a discipline. In mathematics -- *i.e.* geometry (arithmetic, analysis and algebra being left out of consideration) -- "I begin with the definition of my object, such as a triangle or a circle. In metaphysics I can never so begin, and here the definition is so far from being the first thing I know of an object that it is rather almost invariably the last. In mathematics I have no concept whatever of my object before the definition gives rise to it; in metaphysics I have a concept which is already given, though confusedly, and I am to search out the distinct, detailed, and definite concept of the object".[2] The terminological expression of this distinction is the statement that "mathematics achieves all its definitions synthetically; philosophy achieves it analytically":[3] "It is the business of philosophy to analyze concepts which are given as confused, and to make them detailed and definite; but it is the business of mathematics to combine and to compare given concepts of magnitude which are clear and certain in order to see what can be inferred from them."[4]

In the *Notice of the Organization of his Lectures in Winter 1765 to 1766*, Kant again confirmed the difference between mathematics and philosophy using the terms 'analytic' and 'synthetic procedure' to make the distinction: In synthetic procedure "in geometry the simple and the most general are the easiest, in the main science, however, the most difficult. In geometry it has, according to its nature, to be first, in philosophy to be last. In geometry the doctrine starts with definitions, in philosophy one closes with definitions....[5] The concept of time serves as an example for the 'analytic' procedure in philosophy (calling to mind Augustine's remark that he well knows what time is but if someone asks him, he does not know[6]): "If I wished to try to arrive synthetically at a definition of time, what a fortunate accident would have to occur in order for this synthetic concept to be exactly that which fully expresses the idea given to us."[7] According to Kant, some philosophers, if one believes them, claim that such accidents are quite normal. His example: "the philosopher arbitrarily thinks of a substance having the faculty of reason and calls it a mind".[8] Exactly this, however, happens in Leibniz: "Leibniz thought of a simple substance which had only unclear ideas, and he called it a *sleeping monad.* In so doing, he did not explain this monad; rather, he invented it, for the concept of it was not given to him but was created by him."[9] In Kant's opinion Leibniz is thus the prototype of a traditional philosopher whose systematic error consists in confounding or identifying the 'analytic' form of philosophical knowledge with the 'synthetic' form of mathematical knowledge. In this context Kant quotes William Warburton with the remark "that nothing has been more harmful to philosophy than mathematics; or better, there is nothing more harmful than the idea of imitating mathematics as a method of thinking where it cannot possibly be used".[10]

With regard to the argument delineated above, Kant's position is systematically clear, but it is also problematic. Its problematic nature lies not so much in the *use* Kant makes of the distinction between the analytic and the synthetic method, but rather in the *interpretation* which he thus gives of a particular development in the history of ideas. The following is clear: Kant identifies the method of mathematics with the synthetic method which characteristically *constitutes its objects by defining its basic concepts.* In doing so, he follows the ideal of Euclidean geometry, which, right up to modern times, forms the essential paradigm of a logically established *structure* and a well-grounded *presentation* of scientific theories. In this geometry, methods of construction represent the theoretical means for the proof of mathematical existence, the criteria of which are given by the Euclidean postulates. It is not so in philosophy. Philosophy 'acquires' its objects not by construction or definition, but in the form of 'confused' concepts (concepts without distinct parts) which refer to objects given in experience (*e.g.* time). Consequently, Kant believes,[11] philosophy follows the lines of 'Newton's method in natural science' and proceeds 'analytically throughout'[12]; that is, its method is the *explication* of an 'internal experience' described as immediately certain. Yet, in this definition of philosophical method a 'mathematical' orientation is retained to the extent that such an explication ought to proceed by axioms and by deducing conclusions from axioms. Definitions which form the beginning of the mathematical method are, then, *possible* results of such an explication. As 'ultimate material principles of human reason'[13] axioms are, indeed, 'indemonstrable' (example: 'a body is compound'),[14] but are a suitable means of proving other propositions.

Kant fails to notice -- and this is the problematic side of his position -- that the distinction between 'analytic' and 'synthetic', referring here to the form of knowledge of philosophy and mathematics, played no part in the original formation of disciplines and, therefore, of systems of science, but, quite to the contrary, stood for two interrelated aspects of one and the same method. What is meant here is the practice of proof and construction in Greek mathematics, reconstructed methodologically in Pappus of Alexandria.[15] In the framework of the construction of geometrical figures and the proof of geometrical theorems, 'analysis' means a reciprocal method which is opposed to 'synthesis'. 'Analysis' was *analysis of figures*, the whole procedure was divided into three consecutive steps: (1) Axioms, explicit assumptions, theorems already proved and the theorem to be proved were taken as *given*; (2) suitable auxiliary lines were drawn in a figure which represented the fact to be proved in order to determine the searched for assumptions; (3) those auxiliary lines were sought which, by exclusion of the theorem to be proved, permitted a deductive proof from the given parts. The third step establishes the 'synthesis', the application of which is actually made possible by the 'analytic' steps which involve the application of a *heuristic procedure to geometrical figures*. Pappus calls both the 'analytic' part alone and its combination with the 'synthetic' part 'analysis'.

Following the theory of science in Aristotle's *Posterior analytics*, the usual *propositional* conception of analysis arises from an analysis of figures (*i.e.* the geometrical-constructive conception of 'analysis') as a generalization or *logical abstraction*. In reflections on the original procedure aimed at the achievement of generalizations such systems of propositions are also taken into account which are not of a geometrical-constructive kind.[16] In its Aristotelian form and in its application within the framework of Galen's theory of medicine, the Latin Middle Ages take over the propositional conception of analysis (*resolutio*) and synthesis (*compositio*) in the sense of two 'ways to truth'. Here, again, the analytic way from the whole to the parts must be complemented by the synthetic way from the parts to the whole. The meaning of this distinction remains, according to its original geometrical-constructive and propositional conception, a *proof-theoretical* one. This does not change until Paduan Aristotelianism, which returns to scholastic traditions and even further back to the commentaries on Galen rediscovering the older identification of *resolutio* (analytic method) with *demonstratio quia* (proof that [something is as it is]) and *compositio* (synthetic method) with *demonstratio propter quid* (proof why [something is as it is])[17] and making the distinction the central part of a methodology of the empirical sciences. Thus, for Galileo, the *resolutio* serves as an instrument for discovering propositions with which to explain observed phenomena, while the *compositio* leads with the help of the 'analytically' obtained propositions to the formulation of hypotheses which are confirmed by further application of the *resolutio*.[18] Newton finally shifts this distinction decisively in the direction of a methodology of cause and effect: the analytic method has the task of finding causes for observed effects, while the synthetic method argues from observed causes to effects.[19] According to Newton, experimental analysis leads, in the last analysis, to isolated particles and essential properties, from which the phenomena can be deduced synthetically. Thus, in contrast to Galileo's procedure, in which the analytic method is still regarded as an *a priori* clarification of basic concepts and propositions, Newton now subjects analysis to empirical control as well. In other words: despite the fact that the logical structure of the *Principia*[20] still mirrors the constructive ideal of Euclidean geometry and, therefore, the ideal of synthetic methods (within the structure of Euclidean geometry analysis has only one important heuristic function), analysis becomes part of an entirely empirical procedure.

Another step is made in this direction with the so-called 'analytic' theories within post-Newtonian mechanics, starting with Euler's mechanics.[21] In these theories motions are no longer justified by geometric constructions and logical deductions from axioms, but determined by calculating solutions of certain differential equations with respect to certain initial conditions. Causes and effects here appear as measurable quantities, whose dependence on equations defining laws of nature is definitely established. The use of the term 'analytic' to denote, in contrast to the synthetic method, the structure of physical theories starting from basic equations, reflects, apart from Newtonian methodology, the consistent application of the infinitesimal calculus which is called 'analysis' as well as the theory of equations (algebra) which, since Vieta, has also been called 'analysis'.

The history of the concepts 'analytic' and 'synthetic' (sketched here) shows that Kant's comments in the *Prize Essay* not only fail to notice the original meaning of the analysis/synthesis distinction in its geometrical-constructive sense but also take over, far too unhesitatingly, Newton's problematic re-interpretation of the concept of analytic method (together with its elaboration within 'analytic' mechanics in 18th century). When Kant says that the mathematicians have "sometimes defined analytically", but "in every case it has been a mistake",[22] this only shows that he ignores the concept of proof and construction in Greek mathematics, where analysis and synthesis were two well-defined aspects of one and the same method. This fact also weakens Kant's distinction between the 'analytic' form of philosophical knowledge and the 'synthetic' form of mathematical knowledge because, at least in the case of mathematical knowledge, a more differentiated methodology has to be taken into account. It further weakens Kant's claim that Leibniz had identified the 'analytic' form of philosophical knowledge with the 'synthetic' form of mathematical knowledge.

2. *Mathesis universalis*

In contrast to Kant, Leibniz is fully aware of the original relation between analysis and synthesis. At the same time, this relation is the systematic core of the idea of a *mathesis universalis* which, according to Leibniz, should represent the methodological unity of philosophy and science. The point of departure is provided by the corresponding ideas of Descartes which are also based on the systematical order of analysis and synthesis. Descartes had already demanded that all certainty, including philosophical certainty, should imitate the concept of arithmetical and geometrical certainty.[23] To explain the structure of the *mathesis universalis* that he has in mind, Descartes pointed to the analysis of the 'old geometers' and to elementary algebra.[24] This algebra in the form of purely schematical operations of calculation with letters for numerical variables, which, since Vieta, had started to be developed into an independent mathematical theory, is taken as an *algebra speciosa* alongside arithmetic, which, by this time, is the *algebra numerosa*. According to Descartes, it is the task of this algebra, one dealing with the transformation and solution of elementary equations, "to explain of numbers what the old explained of figures".[25] His own formulation of the method of *mathesis universalis* runs as follows: "The whole method consists in the order and arrangement of that on which the gaze of the mind has to concentrate so that we discover a certain truth. We will follow this method exactly if step by step we reduce intricate and obscure propositions to more simple ones and, having succeeded in this, try to raise ourselves, starting from the intuition of the most simple propositions, by the same steps to the understanding of all the remaining ones."[26]

With respect to the conceptual history of analysis and synthesis, the proof-theoretical meaning of this methodological distinction still prevails in Descartes, and this in the original sense of analysis and synthesis of figures. This is also expressed in Descartes's foundation of an analytic geometry which deals with geometrical relations arithmetically in the form of equations by assigning coordinates to points: "just as all of arithmetic is composed of only four or five operations, namely the operations of addition, subtraction, multiplication, division and the extraction of roots,... so too in geometry, in order to transform the lines sought after in such a way that they lead towards the known, one has only to add other lines to them or to subtract other lines from them."[27] The method of philosophy is no exception to this methodological programme. In order to satisfy his own maxim of dealing in philosophy only with those matters which allow a certitude comparable with that of arithmetical and geometrical demonstrations, Descartes, in his replies to the second objections to the *Meditationes*, once again integrates his own procedure into the mathematical method of a *mathesis universalis* as explained in the *Regulae*. The order of this procedure which he wants to be followed in the *Meditationes* 'as accurately as possible'[28] is, according to the formulation chosen here, the following: "that the antecedent ones matters have to be discovered without referring to subsequent ones, and that all subsequent matters have to be arranged in such a way that they are proved by the antecedent alone".[29] He explicitly recalls on this occasion the reciprocal procedures of geometrical analysis and synthesis: "On the one hand, analysis shows the true way by which a thing has been methodically, *a priori* as it were, discovered.... On the other hand, synthesis proceeds by quite a different route. It examines as it were *a posteriori*, though the way of proof is likewise here often even more *a priori* than analysis."[30] According to Descartes, the *Meditationes* follow 'the way of analysis' and not that of synthesis, since synthesis "does not teach the manner in which something has been discovered."[31] Unlike geometry, he continues, with a restriction which calls Kant's comments to mind, "synthesis does not really suit these metaphysical matters."[32] The reason adduced here is that geometry can claim that its 'first concepts' are in accordance with sense perception, a fact which does not hold true of the 'first concepts' of metaphysics.[33] To realize such concepts '*clare et distincte*',[34] however, is the task of an 'analysis' which cannot rest upon constructions 'in empirical intuition' as Kant renders it, though in a different systematic framework. This kind of analysis therefore must proceed otherwise without the usual synthesis of philosophical *disputationes* and systems.[35]

As a matter of fact, Descartes's procedure in the *Meditationes* may be called 'analytic' in a general sense. According to the 'analytic' starting-point from the unknown within the geometrical-constructive and the propositional concept of analysis and synthesis, the following are assumed: the *distinctio* between soul and body;[36] the epistemological reliability of the external world; and the assumption that every certitude is based on false presuppositions. By means of argument an attempt is made to attain certitude in regard to what was previously only apparently known. In working out the scheme, therefore, the demonstration of the *cogito ergo sum* and the demonstration of the existence of God (which aims to guarantee the universal validity of the *regula generalis* of truth as that of *clara et distincta perceptio*[37] which was arrived at along with the demonstration of the *cogito ergo sum*) turn out to be, as Descartes subsequently states, the results of an 'analytic' procedure. Correspondingly, in the sixth meditation the reconstruction of sensuous certitude and the external world is presented as the synthetic counterpart of this analysis.[38] Hence, the argumentative structure of the *Meditationes* corresponds exactly to the presentation which Kant says cannot serve as the method of philosophical knowledge. However, in contrast to Kant's conception of 'analytic' and 'synthetic' in the passage quoted, consideration of the original methodological reciprocity of analysis and synthesis is taken into account by Descartes.

As mentioned before, the same applies to Leibniz, who differs from Descartes in that he establishes the meaning of the analysis/synthesis distinction in the framework of his *theory of concepts.* As far as the idea of a *mathesis universalis* is concerned, it is not the aim of Leibniz's endeavours to arrive at a general theory of philosophical method, as was the case with Descartes to whose 'analytic' orientation he draws attention[39]; what he has in mind is a *formalism* for the representation and for the generation of all knowledge. A kind of *ars combinatoria* is to be undertaken, by which all true propositions are to be generated synthetically from 'simple' basic concepts which are obtained 'analytically'. The structure of a universal language (*lingua universalis*), which serves this intention, follows from the idea of organizing the relation of the words (concepts) of this language to its basic concepts in the same way as the natural numbers are related to prime numbers. The unequivocal reduction of all concepts of language to certain 'simple' basic concepts should imitate the unequivocal analysis of prime numbers. In other words, with Leibniz mathematics becomes the model for all knowledge including philosophical knowledge: "If one could find characters or symbols suited to express all our thoughts as purely and as accurately as arithmetic expresses numbers and geometry expresses lines, one would obviously be able to do with everything which is subject to rational reasoning what one does in arithmetic and in geometry."[40]

The knowledge of mathematical methodology and the influence of Descartes with respect to the paradigmatic character of that methodology for all knowledge reveal their influence on Leibniz's thought in his emphasis on geometrical methods of analysis and synthesis in their original geometrical-constructive and propositional sense. For instance, in a fragment dating from about 1675, he writes: "The method for the solution of any problem is either synthetic or analytic.... It is synthetic or combinatorial if we go through other problems and eventually arrive at our problem. What belongs here is the method of proceeding from simple problems to complex ones. It is analytic if, starting from our problem, we go back far enough to arrive at those pre-suppositions which are sufficient for its solution."[41] At the same time, following the Aristotelian traditions of topics and analytics in Descartes and in the so-called 'logic of Port Royal', aspects which are linked to the distinction between a 'method of discovery' ('*méthode d'invention*') and a 'method of presentation' ('*méthode d'doctrine*')[42] gain importance. Thus, as in the 'logic of Port

Royal', Leibniz sometimes coordinates the analytic method with the sought after *ars iudicandi* and the synthetic method or combinatorial analysis with the sought after *ars inveniendi*.[43] On another occasion he points to the 'inventive' character of both reciprocal methods: "There are two methods, the synthetic method aided by combinatorial science, and the analytic method. Both might point to the origin of the *inventio*. This is, therefore, not the privilege of the analysis. The difference between these methods consists in the fact that combinatorial analysis constitutes a whole science or at least a sequence of theorems and problems, which include what is being looked for. Analysis, however, traces a given problem back to more simple facts."[44]

In this approach emphasizing the 'inventive' character of both methods the concept of *ars inveniendi* functions as a generic term with respect to the concept of *ars combinatoria* or synthesis and *ars analytica* or analysis, while the concept of *ars iudicandi* is equivalent to the concept of *ars demonstrandi* or to logic in its narrow sense. In the sense of *ars combinatoria* or synthesis the *ars inveniendi* is defined as the art of discovering 'correct' questions; in the sense of *ars analytica* or analysis it is defined as the art of discovering 'correct' answers.[45] Where, however, analysis is identified with *ars iudicandi* (it is safe to assume that this was the original conception, subsequently abandoned in favour of the alternative approach[46]), it is defined by two rules, namely (1) 'no concept without explication', (2) 'no proposition without proof'.[47] Also in this sense, that is as a 'method of presentation', analysis remains methodologically dependent on the 'discovering' part, namely the *ars inveniendi*. At any rate, it is beyond question that, for Leibniz, both *artes* belong together - from a historical point of view, *i.e.* in the sense of the original geometrical-constructive and propositional conception of analysis and synthesis, as well as from a systematic point of view, *i.e.* as two elements constitutive for all knowledge.[48]

If one adds to Leibniz's conception the idea of a 'complete encyclopaedia' of knowledge[49] and if one describes such an encyclopaedia as the 'substantial' part of a *scientia generalis* (or *scientia universalis*), in contrast to the *mathesis universalis* as its 'formal' part, Leibniz's methodological considerations result in the following classification:

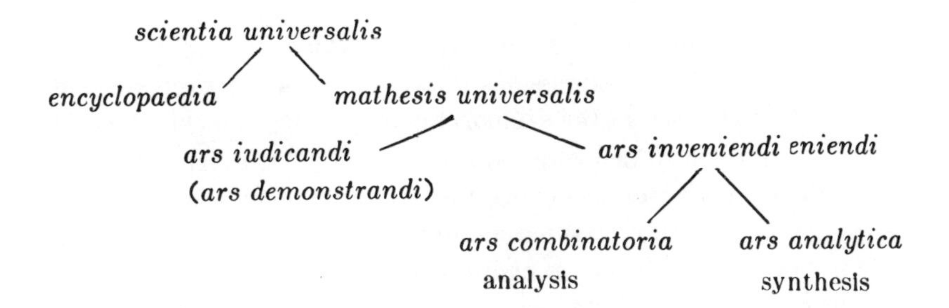

The classification does not obscure the fact that we are dealing here mainly with *ad-hoc* proposals. Nevertheless, the classification does reveal that the conceptions of a *characteristica universalis* or a *lingua universalis* founded in a theory of concepts are coordinated with the *mathesis universalis*. The *ars iudicandi* represents logic in its narrow sense (including a calculus of concepts called '*calculus ratiocinator*', '*calculus universalis*' etc.), the *ars inveniendi* represents logic in its wider sense. Elaborated parts of the programme set forth in the classification are (as paradigms of a *characteristica universalis*) the differential calculus[50] within the framework of mathematics, and different versions of a logical calculus.[51] Here, it is particularly the notion of calculus by means of which Leibniz strives to achieve the methodological unity of all knowledge, including philosophical knowledge.

Leibniz specifies the concept of calculus in the following way: "A calculus or an operation consists in establishing relations; and this is done by the transformation of formulae. The transformations are performed according to certain given rules."[52] Here, the partial concepts of basic figure and basic rule which are constitutive for the concept of calculus and which determine the construction of particular systems of figures are already precisely established; the elements of the calculus used to build up figures represent the sought after 'alphabet of thought' (*alphabetum cogitationum humanarum*).[53] At the same time it is thereby made clear that logical calculi which serve to control in a formal way logical rules of inference are to be understood as parts of a universal language that is also built up like a calculus; this is precisely the *characteristica universalis*. The method of the *characteristica universalis* is articulated as follows: "The *ars characteristica* is the art of forming and ordering symbols in such a way that they represent thoughts and that they are related to one another exactly in the same way as thoughts are related to one another.

An expression is an aggregate of symbols which represent the thing expressed. The rule for expressions is the following: an expression for a thing is to be composed of the symbols for those things out of whose ideas the idea of the thing to be expressed is composed."[54] Corresponding to the aim mentioned above of modelling the construction of concepts on the construction of the truly divisible natural numbers out of prime numbers, the basic symbols of such a *characteristica universalis* are in their composition to be entirely isomorphic with concepts composed of elementary concepts. The language to be constructed may be termed 'characteristic' because (1) its words are produced out of a finite set of symbols (characters) according to well-defined rules of combination; and (2) each symbol 'characterizes' unequivocally the concept determined by it as well as its relations to other concepts.[55]

It is evident that a conception like this one which aims at a 'formal' epistemological unity of philosophy and science within the concept of *mathesis universalis*, cannot easily yield Kant's considerations on the difference between philosophical and mathematical knowledge. The idea of scientific rationality lying behind this conception does not allow a separation between philosophical and scientific rationality. For that reason, it is also right that Leibniz, in contrast to Kant, tries specifically to overcome the difference between philosophy and mathematics. However, this does not happen in the sense criticised by Kant, by confounding or identifying 'analytic' and 'synthetic' forms of knowledge, but, on the contrary, by differentiating the concepts of analysis and synthesis in a historically and systematically plausible way. The idea of Greek mathematics (still realized by Descartes) and the idea of Greek philosophy of science (according to the propositional conception of the analysis/synthesis distinction in the Aristotelian tradition) play their part in this methodological differentiation. Leibniz's intention in making the differentiation was similar to Kant's but the distinctions and results themselves were different.

It is true that when Kant makes his distinction he also makes it clear that what matters to him is the specific form of *philosophical* knowledge, while Leibniz's idea of *mathesis universalis* realized in his concept of calculus characterizes as *mathematical*. It is thus not at all accidental that Kant's critique of Leibniz refers to concepts of the monadology and not to concepts of his logic. Hence, it is not sufficient to meet Kant's critique of Leibniz by presenting distinctions within the conceptual framework of the *mathesis universalis*. The question, rather,is whether substantial parts of Leibniz's philosophy can also be understood in a similar way to the 'formal' parts of the *mathesis universalis*, or, in Kant's terminology: whether the monadology is possible as a *philosophical construction*.

3. The Monadology as Philosophical Construction

With respect to the question of confounding or identifying mathematical and philosophical knowledge, Kant reproaches Leibniz with not having 'explained' but having 'invented' the monad, "for the concept of it was not given to him but was created by him".[56] At first glance, this seems to be exactly the case. Anyone reading the opening paragraphs of the so-called *Monadology* for the first time believes that he has been placed unexpectedly in the unreal world of a metaphysician who, according to the standards of common sense, sees things which do not exist. But this impression is misleading. It is the result of a particular form of presentation, not of the particular way of philosophical research which in this case leads to the monadology. This way is determined by different systematic approaches.

One of these approaches lies in physics, namely in a critique of *physical atomism*. Following contemporary corpuscular theories, Leibniz had himself first adopted the view that elementary units of matter, corpuscles or atoms, exist which differ from each other only in terms of size, form and state of motion.[57] Considerations of the continuum, which are provided with an exact foundation after 1673 by the introduction of infinitesimal methods, bring about the abandonment of this view. The drawing of analogies between the determination of the differential changes of a function or the formation of the sum of differential quantities, on the one hand, and considerations about the structure of the (physical) world on the other, lead to the formation of a new concept: instead of '(material) atoms', reference is now to 'substantial atoms', 'formal atoms' or 'metaphysical points'.[58] In physics the concept of the mass-point is now envisaged. For example, in the calculation of the path of bodies falling within the gravitational field of the earth towards a given point, the bodies occur only as idealizations, *i.e.*, as geometrical points.[59] In *philosophy* the concept of individuation is emphasized, the viewpoint that 'real unities' (real objects) can only be determined by means of conceptual unities: "only metaphysical or substantial points...are exact and real; without them there would be nothing real, since without true unities a composite whole would be impossible."[60] Such unities are described in the *Discourse on Metaphysics* (1686) as *individual substances* (*substances individuelles*) and, from 1696 on, also as *monads.*[61]

Another approach which leads to the monadology is given by the definition of (individual) substance as 'active entity' (*res agens*).[62] Leibniz refers to the characterization of elementary physical units by means of the concept of 'living force' (*vis viva* or *vis activa*),[63] this being a product of Leibniz's critique of Descartes's concept of motion (mv), *i.e.*, the principle of the conservation of motion, which, in the formula mv^2, corresponds to the modern concept of kinetic energy (principle of energy). In addition he recommends use of the concept 'that I have of myself'[64] as the basis for a better understanding of the concept of individual unities. Here, the concept of individual substance is not defined but elucidated by the concept of active entity, and conversely the concept of active entity is not defined but elucidated by the concept of individual substance. At the same time these elucidations indicate further reasons for the *establishment* of corresponding concepts with respect to *given* (physical or pragmatic) facts. The formation of the concept of individual substance or monad proceeds, after all, independently of the characterizations referred to, in the manner of a *logical reconstruction* of the classical concept of substance by means of a *theory of concepts* unfolded within the conceptual framework of the *mathesis universalis.*[65]

According to this *theory of concepts,* which also forms the basis of Leibniz's 'analytic' theory of judgement,[66] individual substances or monads are described by *individual concepts* (*notions individuelles*),[67] constructed as complete concepts, as the (infinite) conjunction of predicates appertaining to the individual they are attributed to. These complete concepts fulfill the Aristotelian[68] conditions for subject-concepts (namely, that they are not able to occur as predicate-concepts). A complete concept is, therefore, a complex predicate that fulfills the conditions of a (complete) description: it is not empty, and there is only one individual to which it can be applied. Ambiguity is avoided precisely by the property of completeness,[69] whereas for the fulfillment of the first condition an existential proof for a 'possible' concept, *i.e.*, for a concept which does not imply logical contradiction, is necessary. Difficulties arise in regard to the function of complete concepts describing predicates of individual substances, from the fact that a complete concept is not, in fact, analysable. Its elements, including, for instance, predicates for the spatial and temporal relations between the substance described by its complete concept and other substances, are infinite;[70] individuality 'includes infinity', as Leibniz says.[71] Complete concepts for *abstract* objects, for instance '*animal rationale*' as a complete concept for *homo*, are an exception, because here the determination of an object is concluded by a definition. In this case, one where the concept is complete but the accompanying description is not infinite, Leibniz speaks of a '*notio plena*' instead of a '*notio completa*'.[72] Furthermore, he occasionally restricts the meaning of a 'complete concept' expressly to *concrete* individual substances (as opposed to abstract objects).[73]

In the cases envisaged by Leibniz, an 'infinitesimal' approximation to a complete concept is possible as a formulation with an arbitrary degree of precision of a complex predicate of an individual substance.[74] What is meant is the following: If S is the complete concept of an individual substance or individual s, and P is a conjunctive part of S, then the statement P(s) to a certain degree divides S into two parts, a known part P and a unknown remainder S_o: $P((S_oP)_s)$. In this case too, the analysis of S is an infinite task when s is a concrete object (or a concrete event) ('*resolutio procedit in infinitum*'[75]), but such a division can be made repeatedly, any number of times, resulting in an increasingly precise knowledge of S. In the context of Leibniz's 'analytical' theory of judgement, this also means an increasingly complete reduction of *contingent* propositions to *identical* propositions (propositions in the form *A est A* or, after substitution of A by AB in the 'virtually identical' case *A est B, AB est B*[76]) and thus to *necessary* propositions.

For my purposes this brief discussion of the concept of individual substance and of the *concept of monad* and its explication within Leibniz's theory of concepts is sufficient. In the framework of this theory of concepts the concept of monad acquires a logical meaning, namely as the unequivocal characterization of individuals (the complete representation of an individual by its complete concept; since such a representation, being infinite, is actually not available, finite representations, *i.e.* finite descriptions, are sufficient here). The *propositions of the theory of monads* (for instance the proposition that every monad represents ['mirrors'] the universe,[77] or the proposition that a pre-established harmony between monads exists)[78] are then partly conceptual consequences, partly (also speculative) supplementations of the concept of monads. From a logical point of view, what is meant by the representation of the universe in every monad is that -- with regard to the description of monads by complete concepts -- statements about any object can also be presented as statements about one and the same object. What is meant by the assertion of a pre-established harmony is the application of this possibility to the (problematical) assumption of an infinite total system which can be represented by a complete concept (the proposition that there are no interactions between monads and that every monad is, therefore, a world of its own, 'without windows',[79] is merely a complementary assertion).

Even the opening paragraphs in the *Monadology* (on the concept of composition) which look highly speculative, become intelligible if one realizes that with the concept of individual substance or monad a *logical subject* takes the place of the *empirical subject*. This has been convincingly worked out by Kuno Lorenz[80] who refers to the representation of the simple (the monad) by the composite (the body), a point clearly stated in the *Monadology*:[81] "because, in general, the complete concept of an object is not available, the body appertaining to it, a 'this-one' ('*Dies-da*') has to replace it symbolically. Alexander, the empirical man Alexander, has to replace the absent complete description (and *a fortiori* the concept) of Alexander, its soul -- he has to represent it symbolically, as Leibniz says."[82] Conversely, the composition of bodies out of monads, 'the entering of the monad into the composites',[83] cannot be taken as a relation of parts; the monad has rather to be understood "as the conceptual articulation of the composite, as 'a principle primarily establishing the unity of its body'."[84] Therefore, it is evident "that the relation between simple substances and composite substances, the relation between monads and bodies, is not given by a synthetic or constructive procedure on the level of objects. Rather it refers to the relation between an object and its linguistic, or to put it more accurately: its conceptual representation -- by means of its name which is identical with a complete description."[85]

We can now observe how in the monadology those aspects are systematically interrelated that for Kant may well be taken as 'mathematical' aspects on the one hand and as 'philosophical' aspects on the other. Nevertheless, the monadology is not (also not in part) a 'synthetical' construction which it would have to be according to Kant's distinction between a philosophical and a mathematical form of knowledge and his identification of the mathematical form with the synthetic method. Two facts speak against this: First, the fact that with his concept of monad Leibniz (logically) reconstructs the classical concept of substance which had been constituted 'analytically' rather than 'synthetically', at least in the Aristotelian form prevailing in the history of ideas. Second, the fact that the constitution of the concept of monad serves in the solution of the problematic unity of the individual in its phenomenal complexity, and that this problem too is, like the concept of the individual, not 'created' synthetically but 'given' (and, no doubt, confusedly given) analytically. In other words: if the concept of monad and if the theory of monads are a *construction*, then, to be sure, this construction serves in a 'philosophical' sense as the clarification of unclear conceptual matters. It is, of course, obvious that here also 'mathematical' elements, among them the older distinction between analysis and synthesis, in its original systematically significant form, play an important part (stimulating attempts, not at all common in philosophical interpretation, to represent the 'structure' of the monadology by means of an 'axiomatic' order),[86] but this does not mean that we are dealing here with a methodologically proscribed step by means of which we depart from the field of philosophical analysis. Again: the concept of construction is not alien to the philosophy of Leibniz, but it would be wrong to say that in this philosophy concepts are just 'thought' or 'created'. They rather emerge from a systematic attempt which definitely includes 'analytic' elements corresponding to what Kant calls the 'givenness' of the objects of philosophical analysis.

4. Construction and Pure Intuition

In the *Transcendental doctrine of method* of the *Critique of Pure Reason*, Kant elaborated the distinction between mathematical and philosophical knowledge drawn in the *Prize Essay*. At the same time he gave this distinction a new systematic basis that focusses attention on the concept of construction. Kant no longer confines himself to the reference to the synthetic method of (Euclidean) geometry in which procedures of construction are the theoretical means for the proof of mathematical existence. Instead he links the concept of (mathematical) construction to a *transcendental analysis* of the concept of (mathematical) intuition. The formation of mathematical concepts is now determined as construction in pure intuition. By making the distinction between empirical and pure intuition as well as by the establishment of the concept of construction in terms of a theory of action, Kant succeeds in gaining important insights into the form of mathematical knowledge which essentially modify the (Euclidean) idea of 'synthetic' mathematics.

Once more the point of departure is the particular epistemological status of mathematics (it "presents the most splendid example of the successful extension of pure reason, without the help of experience")[87] and the distinction between philosophical and mathematical knowledge. The first is defined here as *knowledge gained by reason from concepts*, the second as *knowledge gained by reason from the construction of concepts*.[88] A definition is given of what is meant by the concept of *constructing* concepts, namely "to exhibit *a priori* the intuition which corresponds to the concept."[89] Since this intuition does not refer to empirical intuition, since it has nothing from empirical intuition as its object, "whatever follows from the universal conditions of the construction must be universally valid of the object of the concept thus constructed."[90] This is thoroughly in accordance with the original geometrical-constructive conception of the reciprocal methods of analysis and synthesis ("thus I construct a triangle by representing the object which corresponds to this concept either by imagination alone, in pure intuition, or in accordance therewith also on paper, in empirical intuition -- in both cases completely *a priori*, without having borrowed the pattern from any experience").[91] But now it is reconstructed against the background of Kant's transcendental systematics, in terms of a *theory of action*: in establishing mathematical concepts "we consider only the act whereby we construct the concept";[92] empirical figures (figures 'which we draw') serve only to 'express the concept', and this 'without impairing its universality'.[93] In other words: mathematical concepts do not stand for *figures* of (empirical) intuition, they stand rather for *procedures* of the construction of the figures in (pure) intuition. For Kant this is true of mathematical concepts in general, not only of geometrical concepts.[94]

Beyond this definition of mathematical concepts as constructions in pure intuition (expressed in terms of a theory of action) lies Kant's conception of the schematism, elaborated under the title *The schematism of the pure concepts of understanding*.[95] According to this conception, the procedure 'of providing an image for a concept' is called 'the schema of this concept'.[96] Thus, mathematical rules or methods of construction represent 'transcendental schemata'. These schemata are called 'transcendental' because (according to a somewhat difficult definition) a (transcendental) schema is something which "can never be brought into any image whatsoever. It is simply the pure synthesis, determined by a rule of that unity, in accordance with concepts, to which the category gives expression. It is a transcendental product of imagination, a product that concerns the determination of inner sense in general according to conditions of form (time), in respect of all representations, so far as these representations are to be connected *a priori* in one concept in conformity with the unity of apperception".[97] What is meant here is the following (we are guided by a mathematical example Kant himself uses in our context):[98] the schema of the category 'quantity' is 'number' characterized by two rules of construction

$$n \Rightarrow n|$$

In this arithmetical calculus the first rule signifies an opening rule (a rule without premises) by which the basic figure '|' is constructed, the second rule signifies the transition to another figure: if n, *i.e.*, the premise of the rule, is constructed, then it is permitted to construct n| too, *i.e.*, the conclusion of the rule. The rule-arrow ' $\Rightarrow$ ' signifies here a *practical* 'if-then': whenever an act has been performed resulting in a figure n, another act, by which 'n|' is constructed, shall be permissible according to the rules.

In accordance with the link between 'pure concepts of understanding' (categories) and 'appearances' demanded by Kant for the solution of the problem of applying conceptual means to what is given in sensuous perception,[99] 'number' as a transcendental schema in the sense mentioned above is both 'intellectual' and 'sensuous'.[100] It is sensuous in that the schema of number refers to the structural properties of sequences of acts like beginning, succession and repetition, but not in that it refers 'pictorially' to certain intuitive acts. Thus, number is "a representation which comprises the successive addition of homogeneous units".[101] Its schematic character lies in the repetition of an act starting with a beginning, from which point the pertinent method of construction can make use of different forms of realization -- in the case of the arithmetical calculus (serving here as an example) of sequences of lines or other figures (the identity of figures which here represent ciphers for numbers is guaranteed by two other rules $\Rightarrow | = |$ and $m = n \Rightarrow m| = n|$): schemata like the schema of number therefore represent *characters of action*[102] -- as, by the way, the elements of Leibniz's *characteristica universalis* do also.[103]

With reference to the difference between mathematical and philosophical knowledge this transcendental clarification of the concept of construction leads to the distinction between an *intuitive* use of reason 'from the construction of concepts' and a *discursive* use of reason 'from concepts'.[104] In contrast to the formation of mathematical concepts by construction, "from *a priori* concepts, as employed in discursive knowledge, there can never arise intuitive certainty, that is [demonstrative] evidence, however apodeictically certain the judgment may otherwise be",[105] or, to put it differently: philosophical proof "is always conducted by words".[106] The essential difference between mathematical and philosophical knowledge therefore no longer consists merely in the identification of mathematical knowledge with the synthetic method of (Euclidean) geometry but also in a transcendental reconstruction of the concept of *intuition* or the concept of the *construction* of concepts in pure intuition. True, this reconstruction, in Kant's opinion, confirms his original conception of a fundamental difference between both kinds of knowledge, but it does not restrict this difference only to aspects of the *application* of a method (in the case of mathematics the synthetic method). Furthermore, Kant's analysis of the concept of construction via the (transcendental) concept of schematism provides a suitable means of grounding philosophically mathematical procedures, for instance the geometrical-constructive procedure in its original Greek sense, within the forms of a *transcendental pragmatics*.[107] This fact leads far away from mere classifications, it also for the first time legitimates Kant's claim that the concept of construction functions in the sense of contributing to the system of the sciences, *i.e.*, to the formation of scientific disciplines.

Since the concept of construction is explained by means of the (transcendental) concept of schematism, it is also linked systematically to a 'transcendental analysis' of space and time: constructions are, in Kant's terminology, synthetic procedures for the realizations of forms of judgement (categories) within the intuitive forms of space and time. Hence, space and time, too, as two forms of pure intuition belong to the foundations of mathematics -- the propositions of mathematics are based upon intuitive constructions, *i.e.*, upon the construction of spatial forms or sequences of intuitive figures, upon a 'figurative synthesis' as Kant calls it.[108] Hence, the discovery of a non-conceptual *a priori* organizing the order of 'appearances' leads, particularly in a theory of intuitive space, both to a more specific determination of the concept of construction (as construction in intuitive space) and to the proof that scientific and pre-scientific experience is dependent on the conditions of their spatial production or spatial occurrence. For the first time, Kant attempts to formulate this in a way that is methodologically independent of empirical-physical and formalist mathematical theories. Thus, Kant's concept of space, too, gains a *constructive* character -- with respect to its systematical link with the concept of construction, on the one hand, and with respect to its continuing orientation towards Euclidean geometry, on the other. To this extent, it can be distinguished both from Leibniz's concept of *relational* space (space being determined by possible relations of bodies among each other)[109] and from Newton's concept of *substantial* space.[110]

Therefore, it is the definition of the concept of construction as construction in (pure) intuition ('figurative synthesis') which -- despite the fact that Kant's concepts of space and time in the sense explained have also to be termed *constructive* -- induces Kant to maintain the assertion of the essential difference between mathematical and philosophical knowledge. Moreover, the assertion gains an additional pungency by the determination of the 'limits of pure reason in its transcendental employment'[111] within the framework of the *Critique of Pure Reason*. Against the obstinate attempts in philosophy "to advance beyond the bounds of experience into the enticing regions of the intellectual world" it becomes necessary, according to Kant, "to cut away the last anchor of these fantastic hopes, that is, to show that the pursuit of the mathematical method cannot be of the least advantage in this kind of knowledge (unless it be in exhibiting more plainly the limitations of the method); and that mathematics and philosophy ... are completely different".[112] Once more, Kant emphasizes the fact that 'the exactness of mathematics' rests upon 'definitions, axioms and demonstrations', and that "none of these, in the sense in which they are understood by the mathematician, can be achieved or imitated by the philosopher. I shall show", Kant continues, "that in philosophy the geometrician can by his method build only so many houses of cards, just as in mathematics the employment of a philosophical method results only in mere talk. Indeed it is precisely in knowing its limits that philosophy consists".[113] Leibniz and the programme of a 'reconciliation of geometry and metaphysics'[114] which Kant himself, following Leibniz's theory of monads, had originally advocated,[115] are now definitely placed outside these limits.

5. The Unity of Construction and Reconstruction (A Systematical Outlook)

One of the results of the explication of the concept of construction is the distinction between mathematics as a *constructive* discipline and philosophy as an *analytic* discipline. To this attention has already been drawn. Kant's assertion that philosophy (or metaphysics) hitherto had followed the Leibnizian idea of metaphysics in the spirit of mathematics, is elucidated by the remark that a metaphysics "has never been written"[116] -- it just cannot be written in the spirit of mathematics, according to Kant. Important objections, however, can be made to this assessment -- both with respect to the conceptions of Leibniz and with respect to Kant's own systematic link between the concept of construction and the concepts of pure intuition and the transcendental analytic of space and time. Kant himself actually used the distinction between an 'analytic' and a 'synthetic' method,[117] which in his opinion contributes to the formation of disciplines, in an entirely philosophical way. This distinction permits him to characterize the structure of the *Critique of Pure Reason* as *synthetic*, of the *Prolegomena* as *analytic*.[118] Here, the propositional concept of the original geometrical-constructive distinction between analysis and synthesis is transferred to different forms of philosophical *presentations*.

Such a 'philosophical' employment of a 'mathematical' distinction may, in a certain sense, be 'external' to the philosophical conception which it describes. There are also mathematical reasons which plead for a more 'moderate' interpretation of different forms of philosophical and mathematical knowledge. For, proceeding from the assumption that *reconstruction* is the form of philosophical analysis (including philosophy of science),[119] we find that constructions which, according to Kant, define mathematics are, in a deeper systematic sense, reconstructions which, again according to Kant, define philosophy (in the sense of a philosophical analysis of 'given' concepts). This means that mathematical constructions, from the point of view of scientific foundation, can also be conceived as pragmatic reconstructions.[120] But this exactly is one of the essential results of Kant's analyses within his concept of schematism. Mathematical constructions do have a *pragmatic* foundation. This foundation is included in structures of action. 'Theoretical' objects, natural numbers in Kant's example, are generated through norms for the construction of figures (ciphers) according to rules, and this is achieved by reconstructing a pre-theoretical ('pragmatic') practice to which sequences of acts like beginning, succession and repetition also belong. In this sense it has been said that already in Kant mathematical procedures are founded on transcendental pragmatics. Mathematics and philosophy, if their 'method' were reconstruction, would possess one basis in common.

Thus, Leibniz's arguments in favour of the idea of *philosophical constructions* are not yet exhausted, although, in a strict sense, they can only be formulated within the terminology of Kant's insights. The conceptual powers and weaknesses are well distributed. Kant's distinction between the 'synthetic' form of mathematical knowledge and the 'analytic' form of philosophical knowledge reflects a standpoint which, from the point of view of mathematical methodology, is inferior to Leibniz's approach, but superior to it from the point of view of philosophical methodology, for reasons that do not lie in the distinction between 'analytic' and 'synthetic' knowledge as drawn by Kant. Instead, the reasons are connected with the concept of reconstruction which is not restricted to the level of propositions and conceptual explications but includes a pragmatic level. This makes this concept a constitutive concept both with regard to philosophy or philosophical analysis and with regard to mathematics. A distinction between philosophical analysis and mathematical construction would then lie merely in the dependence of the concept of construction on conditions of pure intuition. Such a dependence does not exist in philosophy, although it is the role of philosophy to explain what the conditions of pure intuition are. In other words: if one emphasizes the systematic unity of construction and analysis within the concept of reconstruction, the positions of Kant and Leibniz do not lie too far apart from each other. This conclusion contrasts strongly with Kant's original view.

Against the background of the history of the concepts of analysis and synthesis, these points also apply to Leibniz's distinction between *ars inveniendi* and *ars iudicandi* to which the distinction between a *constitutive* and a *foundational* meaning of the *synthetic a priori* in Kant, and the distinction between science as *research* and science as *representation* in modern philosophy of science correspond.[121] But this is another important chapter in the modern history of Kant and Leibniz. With respect to Kant's distinction between the synthetic form of mathematical knowledge and the analytic form of philosophical knowledge (in its significance for the system of the sciences) as well as to his transcendental elaboration of the concept of mathematical construction, and with respect to Leibniz's methodological distinction between *ars inveniendi* and *ars iudicandi* as well as to his constructions within the theory of concepts and the theory of monads -- the lesson to be drawn from the chapter we have written here is, among other things, the *philosophical unity of construction and reconstruction.*

Department of Philosophy
University of Konstanz

Notes

1. *Untersuchung über die Deutlichkeit der Grundsätze der natürlichen Theologie und der Moral* [1764], commonly referred to as *Prize Essay,* I. Kant, *Gesammelte Schriften,* published by the *Königlich Preussische Akademie der Wissenschaften,* Berlin 1902ff. (cited hereafter as *Acad.-Ed.*), II, p. 281. I use here the translation of Lewis White Beck (*Immanuel Kant, Critique of Practical Reason And Other Writings in Moral Philosophy,* Chicago 1949).
2. *Prize Essay, Acad.-Ed.* II, p. 283.
3. *Prize Essay, Acad.-Ed.* II, p. 276.
4. *Prize Essay, Acad.-Ed.* II, p. 278.
5. *Acad.-Ed.* II, p. 308.
6. *Prize Essay, Acad.-Ed.* II, p. 283. The reference is to A. Augustinus, *Conf.* XI, 14 (*S. Aureli Augustini Confessionum libri III,* ed. M. Skutella, Leipzig 1934, corr. ed. H. Jürgens and W. Schaub, Stuttgart 1969, p. 275).
7. *Prize Essay, Acad.-Ed.* II, p. 277.
8. *Ibid.*
9. *Ibid.*
10. *Prize Essay, Acad.-Ed.* II, p. 283.
11. *Prize Essay, Acad.-Ed.* II, p. 275, compare p. 286.
12. *Prize Essay, Acad.-Ed.* II, p. 289.
13. *Prize Essay, Acad.-Ed.* II, p. 295.
14. *Ibid.*
15. *Collectiones,* I-III, ed. F. Hultsch, Berlin 1876-1877 (repr. Amsterdam 1965), II, pp. 634ff. See J. Hintikka and U. Remes, *The Method of Analysis. Its Geometrical Origin and Its General Significance,* Dordrecht and Boston 1974, pp. 31ff., also M.S. Mahony, 'Another Look at Greek Geometrical Analysis', *Archive for History of Exact Sciences* 5 (1968/1969), pp. 318-348, and G. Buchdahl, *Metaphysics and the Philosophy of Science. The Classical Origins. Descartes to Kant,* Oxford 1969, pp. 126ff.
16. The 'analytic method' (*μεθοδος ἀναλυτική*) is methodologically classified in the Greek commentators on Aristotle, for instance in Alexander of Aphrodisias (*In*

Aristotelis analyticorum priorum librum commentarium, ed. M. Wallies, Berlin 1883 [*CAG* II/1], pp. 340f.).

17. See A.C. Crombie, *Robert Grosseteste and the Origins of Experimental Science (1100-1700)*, Oxford 1953, pp. 55ff., 61ff., and my 'Changing Concepts of the *a priori*', in: R.E. Butts and J. Hintikka (eds.), *Historical and Philosophical Dimensions of Logic, Methodology and Philosophy of Science (Part Four of the Proceedings of the Fifth International Congress of Logic, Methodology and Philosophy of Science, London, Ontario, Canada - 1975*, Dordrecht and Boston 1977 *(The University of Western Ontario Series in the Philosophy of Science 12)*, pp. 113-128.
18. *Dialogo sopra i due massimi sistemi del mondo* I, *Le Opere di Galileo Galilei*, I-XX, *Edizione Nazionale*, Florence 1890-1909 (repr. 1968), VII, pp. 75f.; compare letter of June 5, 1637 to P. Carcavy, *Ed. Naz* XVII, pp. 88ff. For the concept of 'geometrical' demonstration in Galileo see M. Fehér, 'Galileo and the Demonstrative Ideal of Science', *Studies in History and Philosophy of Science* 13 (1982), pp. 87-110.
19. *Opticks, or a Treatise of the Reflections, Refractions, Inflections & Colours of Light* III, Qu. 31, London ^{4th}1730, ed. I.B. Cohen and D.H.D. Roller, New York 1952, pp. 404f.
20. *Philosophiae Naturalis Principia Mathematica*, London 1687, ^{2nd}1713, ^{3rd}1726.
21. *Mechanica sive motus scientia analytice exposita*, I-II, St. Petersburg 1736.
22. *Prize Essay, Acad.-Ed.* II, p. 277.
23. *Regulae ad directionem ingenii*, *Regula II*, *Oeuvres de Descartes*, I-XII, ed. Ch. Adam and P. Tannery, Paris 1897-1910 (cited hereafter as *Oeuvres*), X. p. 366.
24. *Regula* IV, *Oeuvres* X, p. 373.
25. *Ibid.*
26. *Regula* V, *Oeuvres* X, p. 379.
27. *La géométrie*, *Oeuvres* VI, p. 369; compare *Regula IV*, *Oeuvres* X, p. 393. Descartes's disparaging remark in the *Discours* that the 'analysis of the old' and the 'algebra of the modern' are useless ("the first always being bound to the view

of figures in such a way that one cannot exercise the mind without exhausting the imagination, the second submitting us to the coercion of certain rules and signs in such a way that it has created a confused and dark art which constricts the mind and not a science which improves it", *Discours* II, *Oeuvres* VI, p. 17f.), should not be taken seriously in this connection. It is part of his autobiographical style, which is used to convey the impression that he could learn nothing anywhere (see W. Kamlah, 'Der Anfang der Vernunft bei Descartes autobiographisch and historisch', *Archiv für Geschichte der Philosophie* 43 [1961], pp. 70-84).

28. *Secundae Responsiones*, *Oeuvres* VII, p. 156.
29. *Secundae Responsiones*, *Oeuvres* VII, p. 155.
30. *Secundae Responsiones*, *Oeuvres* VII, pp. 155f.
31. *Secundae Responsiones*, *Oeuvres* VII, p. 156.
32. *Ibid.*
33. *Secundae Responsiones*, *Oeuvres* VII, pp. 156f.
34. *Secundae Responsiones*, *Oeuvres* VII, p. 157.
35. *Ibid.*
36. Compare *Secundae Responsiones*, *Oeuvres* VII, p. 155.
37. Compare *Meditationes de prima philosophia* III, *Oeuvres* VII, p. 35.
38. To elucidate the analytic structure of the *Meditationes*, the reconstruction of the argumentative structure of the '*cogito ergo sum*' which I have given elsewhere, might be helpful (*Neuzeit und Aufklärung. Studien zur Entstehung der neuzeitlichen Wissenschaft und Philosophie*, Berlin and New York 1970, pp. 382ff.), compare in addition, with regard to the analysis -- synthesis distinction, my article on '*mathesis universalis*' mentioned in note* (pp. 601ff.).
39. *Opuscules et fragments inédits de Leibniz*, ed. L. Couturat, Paris 1903 (cited hereafter as *C.*), p. 170.
40. *C.*, p. 155.
41. *Elementa nova matheseos universalis*, *C.*, pp. 350f. Compare letter of March 19, 1678 to H. Conring, *Die philosophischen Schriften von G.W. Leibniz*, I-VII, ed. C.I. Gerhardt, Berlin 1875-1890 (cited hereafter as *Philos. Schr.*, I, pp. 194f. (Synthesis autem est quando a principiis incipiendo

componimus theoremata ac problemata...; Analysis vero est, quando conclusione aliqua data aut problemate proposito, quaerimus ejus principia quibus eam demonstremus aut solvamus).

42. Compare A. Arnauld and P. Nicole, *La logique, ou l'art de penser...*, Paris 1662, repr. under the title *L'art de penser*, ed. B. v. Freytag Löringhoff and H.E. Brekle, Stuttgart-Bad Cannstatt 1965, pp. 303-308.
43. As to the often changing terminology in Leibniz and to the classification of the analytic and the synthetic method within the structure of a *mathesis universalis* compare R. Kauppi, *Über die Leibnizsche Logik. Mit besonderer Berücksichtigung des Problems der Intension und der Extension*, Helsinki 1960 (*Acta Philosophica Fennica*, Fasc. XII), pp. 14ff.
44. *C.*, p. 557.
45. "Duas partes invenio Artis inveniendi, Combinatoriam et Analyticam; Combinatoria consistit in arte inveniendi quaestiones; Analytica in arte inveniendi quaestionum solutiones. Saepe tamen fit ut quaestionum quarundam solutiones, plus habeant Combinatoriae quam analyticе" (*C.*, p. 167 [about 1669]).
46. See L. Couturat, *La logique de Leibniz*, Paris 1901, pp. 177ff.
47. "Analytica seu ars judicandi, mihi quidem videtur duabus fere regulis tota absolvi: (1) Ut nulla vox admittitur, nisi explicata, (2) ut nulla propositio, nisi probata" (*Nova methodus discendae docendaeque jurisprudentiae* [1667], *Sämtliche Schriften und Briefe*, ed. *Preussische Akademie der Wissenschaften*, Darmstadt and Leipzig [later on Berlin and Leipzig] 1923ff. [cited hereafter as *Acad.-Ed.*], 6.1, p. 279).
48. *Discours touchant la méthode de la certitude et de l'art d'inventer pour finir les disputes et pour faire en peu de temps des grands progrès*, *Philos. Schr.* VII, pp. 180, 183.
49. See *Die Leibniz-Handschriften der Königlichen öffentlichen Bibliothek zu Hannover*, ed. E. Bodeman, Hannover 1889, p. 97. For the conceptual vagueness with respect to the relation between the *mathesis universalis* and the projected encyclopaedia see my *Neuzeit und Aufklärung*, pp. 435ff.
50. 'Nova methodus pro maximis et minimis', *Acta Eruditorum* 3

(1684), pp. 467-473 (*Mathematische Schriften*, I-VII, ed. C.I. Gerhardt, Berlin and Halle 1894-1863 [cited hereafter as *Math. Schr.*], V, pp. 220-226). Integrals were introduced two years later: 'De geometria recondita et analysi indivisibilium atque infinitorum', *Acta Eruditorum* 5 (1686), pp. 292-300 (*Math. Schr.* V, pp. 226-233).

51. For a synopsis of the *arithmetical* calculus and different stages of an *algebraic* calculus see my *Neuzeit und Aufklärung*, pp. 440ff., also see K. Dürr, 'Die mathematische Logik von Leibniz', *Studia Philosophica* 7 (1947), pp. 87-102; N. Rescher, 'Leibniz's Interpretation of His Logical Calculi', *The Journal of Symbolic Logic* 19 (1954), pp. 1-13. The pertinent texts are included in the excellent edition by G.H.R. Parkinson: *Leibniz: Logical Papers. A Selection.* Translated and Edited with an Introduction, Oxford 1966.
52. *Philos. Schr.* VII, p. 206.
53. *De organo sive arte magna cogitandi, C.*, p. 430, compare *C.*, pp. 220, 435, also *Philos. Schr.* VII, pp. 185, 199.
54. *Die Leibniz-Handschriften in der Königlichen öffentlichen Bibliothek zu Hannover*, pp. 80f.
55. See Ch. Thiel, 'Leibnizsche Charakteristik', in: *Enzyklopädie Philosophie und Wissenschaftstheorie* II, ed. J. Mittelstrass, Mannheim and Wien and Zürich 1984, pp. 580f..
56. *Prize Essay, Acad.-Ed.* II, p. 277.
57. See *Confessio naturae contra atheistas* [1669], *Acad.-Ed.* 6.1, p. 490.
58. *Système nouveau de la nature et de la communication des substances, Philos. Schr.* IV, p. 482.
59. See 'De linea isochrona, in qua grave sine acceleratione descendent, et de controversia cum Dn. Abbate de Conti', *Acta Eruditorum* 8 (1689), p. 198 (*Math. Schr.* V, p. 237).
60. *Système nouveau..., Philos. Schr.* IV, pp. 483.
61. The expression 'monad', as Leibniz uses it, in all probability comes from the *Kabbala denudata* (I-II, Sulzbach 1677/1684), edited by C. Knorr von Rosenroth and known to Leibniz. In any case it is clear that he had first learned the term from F.M. van Helmont or Anne Conway (see *Nouveaux essais sur l'entendement humain* I1, *Acad.-Ed.* 6.6, p. 72; letter of

August 24, 1697 to T. Burnett, *Philos. Schr.* III, p. 217); compare the entries by van Helmont (*Ad fundamenta Cabbalae...Dialogus*, I/2, p. 309f., and H. More (*Fundamenta philosophiae* 12, I/2, p. 294). Leibniz, in spring and summer 1696, had several discussions in Hanover with van Helmont, whom he had known since 1671. In these discussions he learned about the philosophical and cabbalistical studies of Anne Conway. In 1690 van Helmont published Anne Conway's *Principia philosophiae* (*Opuscula philosophica, quibus continentur Principia philosophiae antiquissimae & recentissimae. Ac Philosophia vulgaris refutata*, Amsterdam 1690); this work appeared two years later, translated back into English, under the title: *The Principles of the Most Ancient and Modern Philosophy Concerning God, Christ, and the Creatures*, London 1692. See C. Merchant, 'The Vitalism of Anne Conway: Its Impact on Leibniz's Concept of the Monad', *Journal of the History of Philosophy* 17 (1979), pp. 255-269, and 'The Vitalism of Francis Mercury van Helmont: Its Influence on Leibniz', *Ambix* 26 (1979), pp. 170-183; R.E. Butts, 'Leibniz' Monads: A Heritage of Gnosticism and a Source of Rational Science', *Canadian Journal of Philosophy* 10 (1980), pp. 47-62.

62. *Principes de la nature et de la grace, fondés en raison* §1, *Philos. Schr.* VI, p. 598; compare *De ipsa natura sive de vi insita actionibusque Creaturarum, pro Dynamicis suis confirmandis illustrandisque* [1698], *Philos. Schr.* IV, p. 509, and *Essais de théodicée sur la bonté de Dieu, la liberté de l'homme et l'origine du mal* [1710], *Philos. Schr.* VI, p. 350.
63. See 'De primae philosophiae emendatione, et de notione substantiae', *Acta Eruditorum* 13 (1694), pp. 11f. (*Philos. Schr.* IV, p. 469); *Système nouveau...*, *Philos. Schr.* IV, p. 478.
64. Letter of July 14, 1686 to A. Arnauld, *Philos. Schr.* II, p. 52.
65. See my 'Substance and Its Concept in Leibniz', in: G.H.R. Parkinson (ed.), *Truth, Knowledge and Reality. Inquiries into the Foundations of Seventeenth Century Rationalism (A Symposium of the Leibniz-Gesellschaft Reading 27-30 July 1979)*, Wiesbaden 1981 (*Studia Leibnitiana Sonderheft* 9), pp. 147-158.

66. See *Philos. Schr.* VII, p. 300.
67. *Discours de métaphysique* §8, *Philos. Schr.* IV, pp. 432f.
68. *Cat.* 5.2a11-13.
69. According to Leibniz's principle of the identity of indiscernibles (*principium identitatis indiscernibilium*) there can be no two individuals who correspond in all their properties. See K. Lorenz, 'Die Begründung des principium identitatis indiscernibilium', in: *Akten des Internationalen Leibniz-Kongresses Hannover, 14-19 November 1966,* III (*Erkenntnistheorie - Logik - Sprachphilosophie - Editionsberichte*), Wiesbaden 1969 (*Studia Leibnitiana Supplementa* III), pp. 149-159.
70. See letter of February 14, 1706 to B. Des Bosses, *Philos. Schr.* II, p. 300.
71. *Nouveaux essais...* III 3, §6, *Acad.-Ed.* 6.6, p. 289.
72. See letter of July 14, 1686 to A. Arnauld, *Philos. Schr.* II, p. 49.
73. "sola...substantia singularis completum habet conceptum", *Fragmente zur Logik*, ed. F. Schmidt, Berlin 1960, p. 479 (= Ph. VII, B IV, 13-14 [ca. 1695]). See G.H.R. Parkinson, *Logic and Reality in Leibniz's Metaphysics*, Oxford 1965, pp. 126f.
74. See *Generales inquisitiones de analysi notionum et veritatum* §74, *C.*, pp. 376f.
75. *Specimen inventorum de admirandis naturae generalis arcanis, Philos. Schr.* VII, p. 309.
76. See *Introductio ad Encyclopaediam arcanam, C.*, p. 513; *Primae veritates, C.*, p. 519.
77. See *Discours de métaphysique* §14, *Philos. Schr.* IV, p. 440; *Monadology* §63, *Philos. Schr.* VI, p. 618.
78. See *Discours de métaphysique* §33, *Philos. Schr.* IV, pp. 458f.; *Monadology* §78, *Philos. Schr.* VI, p. 620.
79. See *Monadology* §7, *Philos. Schr.* VI, p. 607.
80. 'Die Monadologie als Entwurf einer Hermeneutik', in: *Akten des II. Internationalen Leibniz-Kongresses Hannover, 17-22 Juli 1972,* III (*Metaphysik - Ethik - Ästhetik - Monadenlehre*), Wiesbaden 1975 (*Studia Leibnitiana Supplementa* XIV), pp. 317-325.

81. §§61ff., *Philos. Schr.* VI, pp. 617ff.
82. K. Lorenz, *op. cit.*, p. 323.
83. *Monadology* §1, *Philos. Schr.* VI, p. 607.
84. K. Lorenz, *ibid.*
85. K. Lorenz, *op. cit.*, p. 322.
86. See, for instance, B. Russell, *A Critical Exposition of the Philosophy of Leibniz. With an Appendix of Leading Passages*, London 1900, 2nd1937, 1975, pp. 1ff., and G.H.R. Parkinson, *Logic and Reality in Leibniz's Metaphysics*, pp. 182ff.
87. *Critique of Pure Reason* B 740 (I use here Norman Kemp Smith's translation: *Immanuel Kant's Critique of Pure Reason*, London 1929, 1933, 1982).
88. *Critique of Pure Reason* B 741.
89. *Ibid.*
90. *Critique of Pure Reason* B 744.
91. *Critique of Pure Reason* B 741.
92. *Critique of Pure Reason* B 742.
93. *Ibid.*
94. "But mathematics does not only construct magnitudes (*quanta*) as in geometry; it also constructs magnitude as such (*quantitas*), as in algebra. In this it abstracts completely from the properties of the object that is to be thought in terms of such a concept of magnitude", *Critique of Pure Reason* B 745.
95. *Critique of Pure Reason* B 176ff.
96. *Critique of Pure Reason* B 179f.
97. *Critique of Pure Reason* B 181.
98. *Critique of Pure Reason* B 182.
99. *Critique of Pure Reason* B 75f.
100. *Critique of Pure Reason* B 75.
101. *Critique of Pure Reason* B 182.
102. See F. Kambartel, *Erfahrung und Struktur. Bausteine zu einer Kritik des Empirisumus und Formalismus*, Frankfurt 1968, 1976, p. 114.
103. This has been pointed out in detail by F. Kambartel (*op. cit.*, pp. 115ff.) and F. Kaulbach ('Schema, Bild und Modell nach den Voraussetzungen des Kantischen Denkens', *Studium*

Generale 18 [1965], pp. 464-479; *Philosophie der Beschreibung,* Köln and Graz 1968, pp. 228ff., 291ff.).

104. *Critique of Pure Reason* B 747.
105. *Critique of Pure Reason* B 762.
106. *Wiener Logik, Acad.-Ed.* XXIV/2, p. 893.
107. For the justification of this view see my *'Über 'transzendental''*, in: E. Schaper and W. Vossenkuhl (eds.), *Bedingungen der Möglichkeit und transzendentales Denken,* Stuttgart 1984, pp. 158-182.
108. *Critique of Pure Reason* B 152ff. Kant emphasizes here the *constructivist* character of mathematical intuition: "We cannot think a line without *drawing* it in thought, or a circle without *describing* it. We cannot represent the three dimensions of space save by *setting* three lines at right angles to one another from the same point. Even time itself we cannot represent, save in so far as we attend, in the *drawing* of a straight line (which has to serve as the outer figurative representation of time), merely to the act of the synthesis of the manifold whereby we successively determine inner sense, and in so doing attend to the succession of this determination in inner sense. Motion, as an act of the subject (not as a determination of an object), and therefore the synthesis of the manifold in space, first produces the concept of succession -- if we abstract from this manifold and attend solely to the act through which we determine the *inner* sense according to its form" (*Critique of Pure Reason* B 154f.).
109. "spatium...ordo coexistentium phaenomenorum, ut tempus successivorum" (letter of June 16, 1712 to B. Des Bosses, *Philos. Schr.* II, p. 450). Thus, with regard to the isotropy and homogeneity of space in which no direction and no place can be discriminated by mathematical or physical criteria, Leibniz speaks of *abstract* space as "the order of situations, when they are conceived as being possible" (fifth letter to S. Clarke, *Philos. Schr.* VII, p. 415). For the concept of relational space in Leibniz the fact is crucial that space, as a system of relations, possesses the same ideal status already claimed for the concept of physical bodies in the analysis of the continuum, and continued by introduction of the concept

of formal atoms.

110. In Newton's mechanics this concept serves for the definition of absolute movement and also for the definition of inertial movement. The concept had also been originally advocated by Kant (*Von dem ersten Grunde des Unterschiedes der Gegenden im Raume* [1768], *Acad.-Ed.* II, pp. 375-384.
111. *Critique of Pure Reason* B 754.
112. *Ibid.*
113. *Critique of Pure Reason* B 754f.
114. See F. Kaulbach, *Immanuel Kant*, Berlin 1969, p. 59.
115. *Metaphysicae cum geometria iunctae usus in philosophia naturali, cuius specimen I. continet Monadologiam physicam* [1756], *Acad.-Ed.* I, pp. 473-487.
116. *Prize Essay, Acad.-Ed.* II, p. 283.
117. In the 'transcendental doctrine of method', see *Critique of Pure Reason* B 758.
118. *Prolegomena zu einer jeden künftigen Metaphysik, die als Wissenschaft wird auftreten können* [1783], Pref., *Acad.-Ed.* IV, B 263, §§4f., *Acad.-Ed.* IV, pp. 274ff. For a clear methodological distinction between analysis and synthesis in Kant see §111 of his *Logik*: "Die *analytische* Methode ist der synthetischen entgegengesetzt. Jene fängt von dem Bedingten und Begründeten an und geht zu den Principien fort (*a principiatis ad principia*), diese hingegen geht von den Principien zu den Folgen oder vom Einfachen zum Zusammengesetzten. Die erstere könnte man auch die *regressive*, so wie die letztere die *progressive* nennen" (*Acad.-Ed.* IX, p. 149).
119. See my "'Über 'transzendental'" (footnote 107) and my 'Rationale Rekonstruktion der Wissenschaftsgeschichte', in: P. Janich (ed.), *Wissenschaftstheorie und Wissenschaftsforschung*, München 1981, pp. 89-111, 137-148.
120. See F. Kambartel, 'Pragmatic Reconstruction, as Exemplified by an Understanding of Arithmetics', *Communication & Cognition* 13 (1980), pp. 173-182.
121. See K. Lorenz, 'The Concept of Science. Some Remarks on the Methodological Issue 'Construction' versus 'Description' in the Philosophy of Science', in: P. Bieri and R.-P. Horstmann

and L. Krüger (eds.), *Transcendental Arguments and Science. Essays in Epistemology*, Dordrecht and Boston and London 1979, pp. 177-190.

Richard T. W. Arthur

LEIBNIZ'S THEORY OF TIME

1. Introduction

On one level, there is little disagreement concerning the main features of Leibniz's philosophy of time. No one would dispute, for instance, that Leibniz maintained the following three theses:

(i) *Time is relational.* That is, time is not itself a physical entity, but is rather a *relation* or *ordering* of such entities as are not coexistent.

(ii) *Time is ideal.* Being relational, time has no existence apart form the things it relates; it is therefore an *ideal entity.* What exactly Leibniz meant by this is, as we shall see, a matter of dispute: Russell's view that it follows from an ontology which denies the reality of relational facts (such as "a is before b") has recently been forcefully challenged by Ishiguro and others.[1] But whatever its exact meaning, the ideality of time is clearly consonant with Leibniz's belief that *continuity* is a concept that strictly applies only to ideal entities, and that

(iii) *Time is a continuous quantity.*

But if on this level all is relatively well agreed upon, as soon as we take a deeper look at Leibniz's philosophy and attempt to relate these views on time to his metaphysics, in particular his philosophy of monads, there remains little that is not controversial. For in Leibniz's mature view, as expressed in the *Monadology* and elsewhere, all spatial and temporal relations are grounded at the metaphysical level of *actual substances.* That is, they are *phenomenal,* and achieve what reality they possess through being *well founded* in monads and their states. Space is well founded in the mutual *perceptions* of all coexisting monads, and time in the *appetition* of each perception of a monad towards the next. Here the

K. Okruhlik and J. R. Brown (eds.), The Natural Philosophy of Leibniz, 263–313.

term 'perception', which Leibniz appears to use interchangeably with 'state',[2] is misleading inasmuch as monads cannot be said to perceive each other directly: a perception is rather a *representation* of the universe (more or less confusedly) from a given monad's particular point of view. Each such monad, or simple substance is conceived by Leibniz as *continuously active*; but this activity does not consist in the exertion of any influence on other monads, but only in the intrinsic tendency of each state to pass into the next one. The states themselves form an infinite series of states, and this series, just like the mathematical series on which it is modelled, is governed by a law of progression which determines the succession of its terms.

Exactly what is meant by these latter claims remains obscure, despite the enormous effort Leibniz invested in trying to explain them to his contemporaries. But whatever interpretation is given them, there are serious problems in trying to make them consistent in all respects with the commonly accepted features of the theory outlined in the first three theses above. Thus it is difficult to provide any interpretation of the whole theory that is free from contradiction, and this has led commentators to make some rather harsh judgements about the consistency of Leibniz's thought on time. The chief difficulties are as follows.

1. First there is the problem of the *ideality of time*: if time is purely ideal, then how can it apply to actual substances, Leibniz's monads, which he explicitly conceives as temporal successions of states? This difficulty is aggravated if one accepts Bertrand Russell's highly influential interpretation according to whcih the ideality of time is a direct consequence of time's relational nature and the ideality of all relations. As Russell sees it, Leibniz holds that all relations must ultimately be reduced to unary monadic predicates, since these are the only type of predicates which can be said to be completely contained in the subject. Consequently relations, insofar as they are anything apart from such unary predicates, are merely ideal entities which must be superadded by the perceiving mind, and no two monadic states can stand in a temporal relation to each other in actuality.

But this makes it impossible to understand how Leibniz could consistenly conceive time to be founded at the monadic level, or monadic states to succeed each other in time, since on this interpretation time is an ideal relation holding only among perceived phenomena. Russell flatly concludes that the inconsistency is Leibniz's: "time is necessarily presupposed in Leibniz's treatment of substance. That it is denied in the conclusion, is not a triumph, but a contradiction." (RuPL, p. 53).

2. One prospective way of diminishing this inconsistency whilst still preserving Russell's interpreation of time's ideality, is to propose that Leibniz had *two distinct concepts of time*, relational and monadic. Nicholas Rescher and J.E. McGuire, among recent commentators, have both adopted this approach, maintaining that Leibniz's temporal successions of monadic states presuppose an *intrinsic, non-relational monadic time*.[3] For if, as Russell suggests, the relation of mutual concordance between simultaneous monadic states must be reduced to unary properties of the individual monads, then it seems that an individual series of unary temporal properties, or absolute time, must be assumed for every monad. Thus, it is argued, the relational theory is seen ultimately to depend on a myriad of absolute times preprogrammed to keep in step in accordance with the pre-established harmony -- in other words, on this interpretation, Leibniz's relational time presupposes absolute time.

In McGuire's case, a further motivation for this position stems from a consideration of Leibniz's views on continuity. Since the continuity of ideal time precludes its strict applicability to actual substances, the time in which monadic states succeed each other must be "a real, though *discontinuous*, succession of instants" (McGuire, p. 316).

3. Thirdly, it is hard to see how Leibniz can regard time as being *phenomenal* as well as *ideal*. Some authors, following Russell, seem to assume that this amounts to two ways of saying the same thing. Yet in Leibniz's system, real (i.e. well founded) phenomena are distinguished from imaginary phenomena by being founded in actual things as opposed to being merely ideal or possible. Now Leibniz certainly seems to regard phenomenal changes and their order as being founded at the monadic level; consequently, their order cannot be purely ideal. On the basis of such reasoning, McGuire is led to the position that the temporal ordering of phenomena is distinct from ideal time; the former consists in a *discontinuous* succession of changes (McGuire, p. 306 & ff), whilst continuous ideal time functions as a "principle of order" for phenomena (McGuire, p. 314). But this interpretation can hardly be said to relieve Leibniz's views of all contradiction, and indeed McGuire concludes against him that "this exit from the labyrinth of the continuum creates more difficulties than it solves" (McGuire, p. 318).

In this paper I want to counter these imputations of inconsistency by offering a radical reinterpretation of Leibniz's relational theory. The key to this reinterpretation lies in my rejection of the prevailing view that Leibniz conceived relations as obtaining only among phenomena. For I maintain that Leibniz explicitly founded his theory on *relations among monadic states*, and that it is only a combination of some loose translation and the influence of Russell's interpretation that has prevented this from being recognised.

Moreover, I maintain that until one appreciates the true basis of Leibniz's theory in the rational relations among monads, it is impossible to fully appreciate its internal elegance, or its integral connection with his Principle of the Preestablished Harmony. For on my view the Principle of Harmony is the epitome of a relational hypothesis: each state of the universe is a set of monadic states, each of which *is compatible with* all the others, and *contains the ground for* all those which succeed it in time. It is on these two relations of compatibility and ground containment of monadic states, I shall argue, that the whole of Leibniz's theory is grounded.

This still leaves open the problem of the ideality of time. On my interpretation, the ideality of relations pertains only to relations considered as abstract objects, abstracted from all particular relata. This in no way entails their inapplicability to actuals, nor, I shall argue, does it preclude the reality of relations between concrete existents, whether these existents are monadic states or phenomena well founded on them.

The difficulties connected with Leibniz's theory of continuity, on the other hand, are less easily resolved -- though again I believe that the inconsistencies ascribed to Leibniz result more from misinterpretation of his philosophy than contradiction inherent in it. These difficulties were the original focus of this paper, but their resolution is so complex that I was obliged to defer its treatment to a separate paper;[4] here I can only refer to the conclusions reached there without arguing for them. Briefly, I cannot agree with McGuire that Leibniz advocated any discontinuity in change or time, either in phenomenal bodies or in successions of monadic states. On my view, Leibniz conceived any given phenomenal change as infinitely divided into decreasingly smaller, but *contiguous*, discrete parts, and to be truly continuous only when it is considered as an ideal whole formed by abstraction from these parts. Monadic duration is ideal in a different sense, in that its parts, the monadic states, exist "eminently" rather than formally. Although I shall not argue directly for this interpretation here, to the extent that it agrees with my view that Leibniz ascribed ideality only to objects considered in the abstract, it will receive indirect support from my arguments for this below.

But before dealing with the intricacies of Leibniz's views on the ideality of time and abstraction, I want first to consider the prevalent but wholly mistaken claim that his account of time in any case presupposes a monadic time, or private time series for each monad. This will afford us an opportunity to appreciate the full power and coherency of Leibniz's relational theory.

2. The Relational Theory: Against Intra-Monadic Time

Bertrand Russell, although he later became one of the first to recognize its merits,[5] apparently did not understand the *relational theory* of either time or space when he wrote his *A Critical Exposition of the Philosophy of Leibniz* (RuPL) in 1900. For in this work he does not yet appreciate that points and instants can be defined relationally, but rather appears to believe that the fact that they are not themselves relations somehow constituted Leibniz's ground for rejecting the composition of space and time out of them (RuPL, pp. 112-114). Also, in keeping with the conception of relations he attributes to Leibniz, Russell believes that distance, as a relation, "should be analyzed into predicates of the distant terms A and B" (p. 113), and that likewise temporal relations must be analysed into monadic temporal predicates. Thus it is no surprise to find him claiming that "time is necessarily presupposed in Leibniz's treatment of substance" (p. 53), and that "the attempt to infer time from activity thus involves a vicious circle" (p. 52). It *is* a surprise, however, to find these latter claims made in modern expositions of Leibniz's theory of time by authors such as Nicholas Rescher, who is well aware of the modern theory of relations, and of Leibniz's anticipation of its application in the theory of time. According to Rescher,

> Time...has a dual nature for Leibniz. There is the essentially private, intra-monadic time of each individual substance continuing, by appetition, through its transitions from state to state. There is also the public time obtaining throughout the system of monads in general, made possible by the inter-monadic correlations established by the pre-established harmony. Leibniz's standard definition of time as *the order of*

> *non-contemporaneous things* would be vitiated by an obvious circularity if it did not embody a distinction between intra- and inter- monadic time, carrying the latter back to (i.e. well-founding it within) the former. (RePl, p. 92).

Similarly, J.E. McGuire objects:

> If time is an ideal notion, it does not apply to the actual. But actual substances have expressed states, are expressing states, and will express states. Moreover, as they are states of one and the same individual substance, that substance is programmed to unfold a unique history. But such action implies not only some notion of continuity but some conception of monadic time. (McGuire, p. 312).

But, disregarding for now the question of ideality, *does* the activity of monads in time imply that they have their own *private* time? Certainly, it must be granted that monadic states precede and succeed each other in time, since Leibniz explicitly claims this, as we shall see shortly. But the implication of the above criticisms is that unless the time in which they precede and succeed each other is *different* from the time "obtaining throughout the system of monads in general" (Rescher), or from phenomenal time (McGuire), there will be a vicious circularity. This echoes Russell's criticism that in Leibniz's treatment of substance an individual state of a monad can only be identified as one occurring *at a given instant*, so that time cannot be defined in terms of relations among monadic states without circularity.[6]

As I have already suggested, I think these charges betray a serious misunderstanding of the relational theory of time, not just in its post-Russellian manifestation, but in Leibniz's own version. The best way to demonstrate this is by putting the theory in mathematical form. By stating Leibniz's axioms and definitions in set-theoretic notation, it can be shown how both *instants* and *time* can be constructed on this basis without in any way presupposing that each monad has its own intra-monadic time. I have relegated the details of this formulation to an appendix so as not to interrupt the flow of the argument. Here I shall attempt to convey the cogency of the theory by means of a sketch.

Leibniz gives the most complete account of his mature theory of time in the *Initia Rerum Mathematicarum Metaphysica* (hereafter abbreviated as the *Initia Rerum*), which he penned in 1715, the last full year of his life. There he explicitly defines simultaneous states of monads as those which do not "involve opposite states"; temporal precedence, on the other hand, is defined in terms of one state "involving the reason for" another:

> *If many states of things are assumed to exist, none of which involves its opposite, they are said to exist simultaneously.* thus we deny that the events [*quae...facta sunt*] of last year are simultaneous with those of this year, since they involve opposite states of the same thing.
>
> *If one of two states which are not simultaneous involves the reason for the other, the former is held to be the earlier,* and the latter to be the *later.* My earlier state involves the reason for the existence of my later state. (*Initia Rerum*, GM. VII. 18)

It is tempting to try to capture the sense of these remarks by simply defining one state as temporally preceding another if and only if it precedes it in the order of reasons (i.e. includes the reason for it). But this approach would only give us an intrinsic temporal order for each monad, and simultaneity of states of different monads would then have to be introduced in terms of these orders, thus falling afoul of Russell's criticism.

But Leibniz introduces simultaneity first, defining it in terms of the *non-opposition* of monadic states: *one state is simultaneous with another if and only if it is not incompatible with it.* The beauty of this strategy is that we can now determine which states occur at the same time without any recourse to instants -- provided, of course, that we can give a consistent rationale for the compatibility of states which does not presuppose their temporal relations.

This can be achieved as follows. First we assume that any state which involves the reason for another is incompatible with it, so that all the states of any one monadic series are mutually incompatible. Next we define simultaneous states as those which are not incompatible. But in order for this strategy to give us unique and well-defined classes of simultaneous states, we need some axiom to connect compatible states of different monads. For what we cannot rule out with the axioms we have so far is the situation where one state is compatible with (and thus simultaneous with) a second, and the second with a third, even though the third is incompatible with the first. In other words, it is possible for simultaneity to be non-transitive. Now it is precisely this possibility that Leibniz rules out by positing the "connection of all things" in the continuation of the passage in the *Initia Rerum* quoted above:

> *If one of two states which are not simultaneous involves the reason for the other, the former is held to be the earlier,* and the latter to be the *later.* My earlier state involves the reason for the existence of my later state. And since, because of the connection of all things, my earlier state involves the earlier state of the other things as well, it also involves the reason for the later state of these other things, so that my earlier state is in fact earlier than their later state too. And therefore *whatever exists is either simultaneous with, earlier than, or later than some other given existent.* (*Initia Rerum*, GM. VII. 18)

Interpreting one state's "involving" another as meaning they are compatible, and hence simultaneous, this becomes Axiom 4: if a is simultaneous with b, and b involves the reason for c, then a also involves the reason for c. What this axiom guarantees is that (so long as there exists a plurality of monads and their states), any of these states is a member of a unique class of states all of which are compatible with it. Now the quality which all these states have in common is that they are simultaneous, i.e. occur at the same time. So we may define an *instant* by abstracting from the entire class of simultaneous states what they have in common, namely their membership in this class: to occur at an instant is to be a member of such a simultaneity class. Moreover, the axiom of connection ensures that the orderings of states of every individual monad correspond with each other, so that there is just one order of reasons governing the whole universe. Correspondingly, there is just one temporal ordering, and, as Leibniz claims, this is a *total ordering*: any given state is either simultaneous with, earlier than, or later than any other possible state.

As it stands, however, Leibniz's time concept still lacks two essential ingredients: some notion of *continuity*, and a way of defining temporal *magnitude*. In fact, it was on essentially these grounds that his theory of time was attacked as inadequate by his opponents, the Newtonians, who conceived time neo-Platonically as a Quantity, or self-existent form.[10] In order for time to be a quantity, argued their spokesman Samuel Clarke, it must be something more than just an order:

> The order of things succeeding each other in time is not time itself, for they may succeed each other faster or slower in the

> same order of succession but not in the same time. (Clarke to Leibniz, IV, #41: G.VII)

Leibniz replied by denying that the order of succession *could* remain the same if the time were greater or less:

> For if the time is greater, there will be more successive and like states interposed, and if it is smaller, there will be fewer, since there is no vacuum in times, nor, so to speak, any condensation or penetration, any more than there is in places. (To Clarke, V, #105: G. VII. 415)

This passage has often been interpreted as evidence that at bottom Leibniz held time to be *atomistic.*[11] For if the number of states in any given time is countable, and if there are no temporal gaps or "vacua" between successive states, it appears to follow that each state must be of a certain discrete duration. At any rate, the states cannot form a continuum in the modern (Cantorean) sense, since the number of elements in a Cantorean continuum is nondenumerable, and the notion of a *next* element is undefined. So it appears that Leibniz's references to "atoms of substance" and "unities of duration" (e.g. in his correspondence with de Volder, 11/10/1705: G. II. 278-9) are to be understood as signifying a commitment to some kind of spatial and temporal atomism.

This is not Leibniz's position, however. For despite his frequent references to them as "unities", he is adamant that monads and their states *have no spatial or temporal extension,* so that space and time *cannot be composed out of them.* This point is an extremely subtle one, and I cannot do it full justice here (though I have tried to do so in my paper on Leibniz's theory of continuity).[12] The basic idea is that although monads and their states are "discrete" in the sense of being "qualitatively distinct", and are therefore countable, *they have no measures,* and so cannot be "composed" into continuous lengths or times. In fact Leibniz was well aware that the denseness property of the continuum entailed that there was no "next point" in a continuum of points,[13] and that this definitely ruled out the composition of a continuum out of points or atoms.[14] In his view, the indefinite divisibility of the continuum evidenced its "incompleteness": a truly continuous thing has no determinate parts, but only an indefinite possibility of division. Since nothing actual can exist without determinate parts, continuity must apply only to things considered abstractly. But just as an irrational number can be *resolved* into integers (and thus unities) by representing it by means of an infinite series, so any continuous phenomenon is resolvable into monadic unities, and is based on a particular infinite series of states.[15]

But the question remains how the number of monadic states can have any bearing on the quantity of a continuous interval of time. It is no good saying that there are more states in a greater time if there is no bridge from the denumerable quantity of states to the continuous quantity of time. The bridging of the gap is precisely what Leibniz sought to achieve in the theory of spatial and temporal magnitude which he sketches in the *Initia Rerum.* There he attempts to define magnitude by the application of Extremal Principles to the order of perceptions: the interval between two states is the simplest order of succession between them, and this may be determined by the application of extremal principles to the phenomenal structures arising from this order:

> In each of the two orders (time and space), elements are judged to be *nearer to or farther from each other* according as *more* or *fewer* are required for the order between them to be understood. Hence two points are nearer for which the interposed points maximally determined from them produce something simpler. Such a maximally determined interposition is the simplest path from one to another, i.e. the shortest and at the same time most uniform path, namely, the straight line, which is less interjected between nearer points. (*Initia Rerum,* GM. VII. 18)

As Leibniz explains later in the *Initia Rerum,* the application of these principles of optimization and economy to phenomena has its foundation in the monads themselves, the nature of a thing being such that the order between any two of its states or perceptions is the simplest[16]:

> There is, moreover, a certain order in the transition of our perceptions when it passes from one to another through others between them....But since this order can vary in infinite ways, there must necessarily be one simplest order, which would in fact be that order which proceeds according to the thing's own nature through determinate intermediate perceptions, i.e. through those which are related as simply as possible to each of the two extrema....And this simplest order is the shortest path from one to the other, whose magnitude is called *distance.* (*Initia Rerum,* GM. VII. 25)

The details of this theory of spatial and temporal magnitude, however, remain somewhat obscure; and although it would be a fascinating exercise to investigate to what extent the theory could be explicated with the methods of modern topology (which is itself a descendant of Leibniz's *Analysis Situs*) and measure theory, the complexity of this topic and its entanglement with the web of Leibniz's thought on continuity preclude its proper treatment here. Accordingly I shall not discuss the metrical aspect of Leibniz's theory any further in this paper.

It should be remarked here that my interpretation of the passages I have quoted form the *Initia Rerum* in the above exposition is not the usual one. Indeed, these passages are usually interpreted as referring exclusively to *phenomena*, the relation of "involving the reason for" (or "ground containing") being interpreted as the *causal* relation. Part of the reason for this, I believe, is loose translation. Phillip Wiener, for example, in the first complete English translation of the *Initia Rerum*, renders Leibniz's "states of things" (*rerum status*) first as "concrete circumstances", then as "elements", and then as "states of existence", all in the same passage. This effectively disguises the fact that "my earlier state" is an instance of the "states of things" previously referred to, and leaves the impression that the "concrete circumstances" are phenomenal. Likewise, Hermann Weyl's (English) translation of the relation "includes the reason for" as "comprehends the cause of" has led modern authors such as van Fraassen to conclude that "according to Leibniz the various circumstances or states of affairs are related to each other as cause to effect" (vanF, p. 38).

I will come back to this interpretation -- and indeed to the whole question of time as an order of phenomena -- in section 4 below. But for now, irrespective of whether the *Initia Rerum* is properly to be interpreted as concerning monadic states or phenomena (or both), a comparison with what he wrote elsewhere makes it clear that Leibniz *did* intend his relational analysis to apply to the series of states of monads. In his correspondence with de Volder, for instance, Leibniz explains the involvement of all the future states of a monad in a given state by reference to the *law of the series*. Indeed it is the implicit involvement of the whole series in any one state that makes for the *self-identity* of the individual monad:

> The succeeding substance is held to be the same when the same law of the series, or of continuous simple transition, persists; which is what produces our belief that the subject of

> change, or monad, is the same. That there should be such a persistent law, which involves the future states of that which we conceive to be the same, is exactly what I say constitutes it as the same substance. (To de Volder, 21/1/1704: G. II. 264)

Leibniz is even more explicit in his controversy with Samuel Clarke, in the course of which he explains that this is the very meaning of the Principle of Pre-established Harmony, given that each monadic state is a representation of the universe from its own particular point of view:

> Since the nature of every simple substance, Soul, or true Monad, is such that its following state is a consequence of the preceding one; it is there that the whole cause of the *harmony* is to be found. For God has only to make a simple substance be once and from the beginning a *representation of the universe*, according to its point of view: since from this alone it follows that it will be so perpetually, and that all simple substances will always have a harmony among themselves, because they always represent the same universe. (To Clarke, V. #91: G. VII. 412)

In other words, the fact that a state is a representation of the same universe guarantees that there exists a unique class of possible monadic states which correspond with it, or are in harmony with it, and are thereby simultaneous with it. And since one state involves the reason for all states subsequent to it in the same monadic series, it thereby contains the ground for all future monadic states of the same universe. This is an unambiguous statement of the axiom of connection for all the monadic states of any given universe.[17]

Moreover, the axiom of connection also provides the means for constructing Leibniz's concepts of *compossibility* and *possible world*. For Leibniz intended this axiom to mean not just that an actually occurring state represents the whole existing universe from its own point of view, but that any possible state so represents the whole of a given universe of possibles. After all, there are innumerable possible states of monads that are compatible with any given state, but in order for them to coexist in the same world they must belong to monadic series which are mutually compatible everywhere. That is, the simple substances of which they are states must be in harmony from beginning to end. Any two such monads are *compossible*: to any state of one, there corresponds a unique compatible state of the other. And a complete set of compossible monads is what Leibniz calls a "universe", or *possible world.*

Now of all these possible worlds, existing in the mind of God, only one is brought into existence *in re*, namely the one containing the maximum degree of perfection. This is the *actual universe,* W_a:

> [All] possibles are not compossible. Thus the Universe is only the collection of a certain order of compossibles; and the actual Universe is the collection of all the compossibles which exist, that is to say, those which form the richest composite. And since there are different combinations of possibilities, some of them better than others, there are many possible universes, each collection of compossibles making up one of them.[19] (To Louis Bourguet, Dec. 1714: G. III. 573)

Evidently, then, all the relations on which Leibniz erects his theory are relations among states of monads in any of these possible universes. But from this it follows that time, as Leibniz construes it, does not just pertain to the actual universe, but is the order of succession of *any possible universe.* This is an important point, and I shall come back to it in the next section.

Returning to our initial problem, I think the above analysis vindicates the consistency of Leibniz's theory against the charges of vicious circularity. Certainly there is no circularity in defining temporal succession in terms of the logical relation of reason inclusion, or of simultaneity in terms of logical compatibility. Thus the state comprising my being here in London is logically incompatible with that comprising my being in Toronto, so that if both states occur in the same universe we can deduce that they are non-simultaneous. On the other hand, whether my current states include the reason for my meeting Thomas Kuhn on Saturday coming, I do not know, and according to Leibniz only an infinite analysis could decide. Certainly a state comprising my meeting him is a *possible* monadic state, unlike my being in two places at once; but we need to know whether it is *compossible* with what already exists, and it is this that requires the infinite analysis (and renders the fact *contingent* rather than necessary).

Granted, such a rationalistic view of the world is unlikely to be accepted today, but this is not germane to the charge of circularity. For once Leibniz's axioms are accepted, it follows from the fact that one state includes the reason for another that it precedes it in time, and that any two states which are compatible states of compossible monadic series are thereby simultaneous. Thus Russell is wrong in claiming that time is presupposed in Leibniz's treatment of substance through the reference to future states, and that one state can only be defined through reference to an instant of some intra-monadic time. Similarly, since instants of time are only defined in terms of the compatibility of the states of *different* possible monads, there are no intra-monadic instants, and therefore, *contra* Rescher, there is no "private, intra-monadic time", despite the fact that each monadic state does indeed occur at some given time (is a member of some compatibility class of possible monadic states).

Nevertheless, there is a kernel of truth in Russell's and Rescher's charges of circularity: for there is a kind of circularity in the mutual dependence of time and change, though it is not in my view in any way vicious, but rather essential. For as we have expounded it so far, Leibniz's theory of time is a *static theory of time*, very much of a piece with the "becomingless theory" advocated in recent years by Adolf Grunbaum.[19] We have set up a one-one correspondence between the temporal order of states and their logical ordering by the relation of reason inclusion; but we do not yet have any principle of change or becoming, whereby a monad, having attained a given state, passes into the next state in the series.[20] Without such a principle, of course, there would never *be* any series of states which could be temporally ordered (this is the fallacy of Grunbaum's attempt to distinguish becoming from temporal change, with the latter viewed solely as "difference" between states in the temporal order). Hence a circularity exists in that there would be no time without becoming, and yet all things which become do so in the order we call time.

Leibniz explicitly postulates a principle of becoming for his monads: it is precisely what he calls *appetition*. In the *Monadology* he states:

> 10. I also take it for granted that every created being, and consequently the created Monad too, is subject to change, and even that this change is continual in each....
>
> 11. It folows from what has just been said that the natural changes of monads come from an *internal principle*, since an external cause cannot influence its interior [*ne sauroit influer dans son interieur*]...
>
> 15. The action of the internal principle which produces change or passage from one perception to another may be called *Appetition*. (G. VI. 608)

The postulation of such a principle of temporal change, or becoming, is thus in a sense equivalent to the postulation of time. But once becoming has thus been postulated, the succession of states can truly be said to be *temporal* -- *despite* the fact that the theory posits a one-one correspondence between the temporal series of states and the atemporal logical order of reason inclusion. As Leibniz defends himself against de Volder:

> You say that in a series (such as a series of numbers) nothing is conceived of as [temporally] successive. What of it? I do not say that any series is a [temporal] succession, but that a [temporal] succession is a series, and has this in common with other series, that the law of the series shows what point should be arrived at in continuing the progression, or in other words, that given the initial term and the law of progression, the terms should proceed in an order, whether this be an order or priority in nature only, or also one in time. (To de Volder, 21/1/1704: G. II. 263)

It should be noted in passing that the above discussion makes perfectly transparent the meaning of Leibniz's much-quoted reference to the "priority in nature" of one instant over those following it in his letter to Bourguet:

> I admit however that there is this difference between instants and points: one point of the universe has no advantage of priority over another, whereas the preceding instant always has the advantage of priority, not only in time, but also in nature, over the following instant. (To Bourguet, 5/8/1715: G. III. 582-583)

This reference to a "priority in nature" has caused some rather extravagant misinterpretations of Leibniz's theory of time. Rescher (RePL, p. 101) makes it the basis for his claim that "There is a total lack of homogeneity in time, according to Leibniz", in blatant contradiction to Leibniz's own teaching;[21] and McGuire seems to think it involves the notion of "non-temporal instants" making up a "real, though *discontinuous* succession" (McGuire, p. 316). But all that is involved is the fact that instants are intrinsically ordered by G (priority in nature) as well as by $<t$ (priority in time). [Definition 3(a) + preceding axioms].

In conclusion, then, it is evident that Leibniz's relational theory of time does not presuppose individual time series for each monad, or "intra-monadic times", but rather makes them wholly redundant. As Russell said, (although ironically he intended this as criticism of Leibniz), "There is, in short, *one* time, not as many times as there are substances" (RuPL, p. 52).

3. The Ideality of Time and Leibniz's Treatment of Relations

There can be little doubt that one of the main motivations for the erroneous belief that monads have to be equipped with their own private times lies in Russell's thesis that Leibniz held that all relations must be reduced to unary monadic predicates -- the thesis of the Reducibility of Relations. According to Russell,

> Leibniz is forced, in order to maintain the subject-predicate doctrine, to the Kantian theory that relations, though veritable, are the work of the mind (p. 14)...[R]elations must always be reduced to attributes of the related terms (p. 122). (RuPL)

In this he was followed by Louis Couturat, and, more recently, by Rescher, who restricted the application of the thesis to relations obtaining among individual substances (RePL, p. 74). As a consequence of the thesis, both Russell and Rescher reason that time is ideal *because* it is relational, relations being only the ideal manifestations of the unary monadic predicates, which alone are real, and to which they must be reduced. Thus Rescher:

> Space and time are ideal, or rather are phenomena, space because it is nothing but the order or relation of simultaneous existents, and time since it is relational, and involves the labyrinth of the continuum (pp. 96-97) ...[D]istance, being a relation, has no place in the monadic realm (p. 101). (RePL)

This is not the place to attempt a comprehensive refutation of the reducibility thesis, and in any case I believe something very close to this has been achieved already by Ishiguro (in Ish). Certainly the texts do not provide unambiguous support for the thesis. As Ishiguro points out, Leibniz's concept of 'predicate' is more grammatical than metaphysical, and cannot simply be assumed to exclude what we would now call "open sentences with one free variable" -- i.e. propositions of the form xRa or (∃y)(xRy) -- whether these express *intrinsic* relational properties, such as "contains the ground for state a", or *extrinsic* ones, such as "is sitting next to the Ayatollah" (Ish, Chs. 5 & 6, esp. pp. 94-110). But for my purposes here, I believe the most germane criticism of the thesis of the reducibility of relations is that when it is applied as an interpretation of Leibniz's treatment of relations in his theory of time, it appears to be patently self-refuting.

It is a credit to their honesty that both Russell and Rescher recognize the contradictions which follow from applying this thesis to Leibniz's theory of time, but not to their credit, I think, that they blamed Leibniz for the inconsistency. Thus Russell acknowledges that Leibniz writes about relations among monads in an acutely non-reductionistic fashion. For instance, in the "Principles of Nature and Grace" (1714), Leibniz states that:

> [T]he simplicity of a substance does not preclude the multiplicity of the modifications which must be found together in this same simple substance; and these modifications must consist in the variety of relations [*rapports*] it has to the things outside it....Everywhere there are simple substances, actually separated from each other by their own actions, which continually change their relations. (G. VI. 598)

Commenting on the latter passage, Russell says that

> the important point is, that the relations, being among monads, not between the various perceptions of one monad, would be irreducible relations, not pairs of adjectives of monads. In the case of simultaneity this is peculiarly obvious....(RuPl, p. 130)

But the relations which are said to change here are relations of similarity and correspondence between different monadic representations or points of view -- that is, relations between *monadic states*, just as are the relations of reason inclusion, incompatibility, simultaneity etc. dealt with above. There therefore seems to be little justification for Russell's interpretation of this passage as evidence for an "objective" system of unmediated relations among monads, as opposed to the "subjective" system of relations among their perceptions: the dichotomy which Russell attempts to foist onto Leibniz fails to gain a foothold. Nevertheless, the important point, as he says, is that relations such as simultaneity *have* to be irreducible, otherwise no sense could be made of Leibniz's theory of time without imputing to it a "fatally vicious circle".

Similarly, Rescher concludes his account of Leibniz's doctrine of relations by conceding "a serious difficulty with Leibniz's relation-reducibility thesis":

> It is clear that only genuinely relational properties -- properties not representable by genuine predicates -- can underwrite the incompatibility of alternatively possible substances. (RePL, p. 79)

-- a difficulty which undermines the relation-reducibility thesis in this context just as surely as did Russell's difficulty about simultaneity.[22] But how credible is it that Leibniz would define time and compossibility in terms of relations among monadic states, only to deny that such relations ultimately exist? Are we really to believe that the same person who conceived the universe as a temporal succession of a set of harmoniously related states, each representing it according to its situation in the order, had no place for relations in his ontology?

I submit that the alternative interpretation of the ideality of relations offered by Hide Ishiguro is the correct one: namely that it is *relations considered as abstract entities*, or things in their own right, that Leibniz held to be ideal. Thus space and time are ideal, in the first instance, because they are *abstractions* from extended and enduring substances. At any rate this is the principal reason for their ideality, though we will see that there are some connected subsidiary reasons as well. Thus Leibniz argues against Malebranche in the "Conversation between Philarete and Ariste" (Ca.1711):

> I deny that extension is a concrete term, since it is an abstraction from the extended. (G. VI. 582)

But although they are abstractions, extension and duration may nevertheless also be considered as attributes of things: one can sensibly talk about the extension or duration of a given object or process. It is not the same, though, with extension or duration *in general*;[23] and it is when they are considered in this purely abstract sense that we call them space and time:

> Duration and extension are attributes of things, but time and space are considered to be something outside of things, and serve to measure them. ("Conv. b. Phil. & Arist.": G. VI. 584)

Similarly, he urges Clarke to recognize this distinction:

> Things keep their extension, but they do not always keep their space. Each thing has its own extension, its own duration; but it does not have its own time, and does not keep its own

space. (To Clarke, V, #46: G.VII.399)

Actually, this is a traditional distinction, and not one of Leibniz's own innovations. Indeed, he is here appealing to the Newtonians in a common language -- or at least the one which comprised both the Newtonian and Leibnizian starting points -- that of the Cartesians. For even Descartes upheld the distinction between "the duration of the enduring thing" and "time considered in the abstract", parallelling that between portions of extended matter, on the one hand, and "generic extension" or space on the other. In Newton these become his *relative* times and spaces, as opposed to the mathematical *absolute* time and space. What is original with Leibniz is the typically precise way in which he expounds this distinction, and its close connection with the other principles of his system -- especially the Identity of Indiscernibles.

Contrary to a superficial reading of his philosophy, Descartes himself was of course well able to distinguish between mathematical objects, which could exist simply as ideas, and their manifestations in actuality: although extension is the essence of matter, the two cannot simply be identified. Thus there is a distinction between the "true and real beings" of mathematics -- extension, purely geometric lines and surfaces etc. -- and "actual and existent beings", such as souls and matter, which need a *force* to bring them into existence and to conserve them in existence as long as they endure. This force Descartes calls *conatus*, literally an effort or striving; it is conceived as an instantaneous *action* or tendency toward motion, and it is a necessary condition for the continued existence of the universe that God should exert it at any instant of the duration of the universe in the same degree that he exerted it in creating the universe. (Hence the conservation of the total quantity of motion in the universe, according to Descartes.) Leibniz, knowing Descartes primarily through his followers (who tended to ignore this distinction), independently rediscovers the necessity of force for existence and actuality, and urges this as demonstrating the inadequacy of a purely geometric physics, and the need for a dynamics. Thus, for Leibniz too, in order for something to *exist* or be *actual* it must be sustained throughout its duration by a force; also it must have a unique and determinate identity throughout this duration, a point we shall return to below. In fact this trend of thought accounts for a certain ambiguity in Leibniz's use of the term 'actual'. For in one of its connotations, the *actual* is opposed by him to the *phenomenal*, actuals being the true existents from which phenomena result. But we have also seen a second sense in which he uses the term, when he distinguishes the *actual* world -- the one that *exists* -- from the merely *possible* ones, which exist merely as ideas in the mind of God. It is this second sense which corresponds to Descartes' usage of the word: if things are to be actual in this sense, then they require a force to sustain them in existence for their entire duration. It is in this sense, too, that Leibniz can talk of "actual phenomena", for instance, these being phenomena that do in fact occur (as a result of actual monadic concordances),[24] as opposed to phenomena that might have occured (as results of possible monadic concordances); or the "actually infinite division" of phenomenal bodies, as opposed to their purely ideal divisibility. Phenomenal *things*, of course, just like individual substances, need a force to sustain them in existence; though, unlike the *primitive forces* sustaining the monads through eternity, these *derivative forces* are contingent and ephemeral.

By this criterion, then, time is already ideal, not being the sort of thing that could be said to exist for a duration or be sustained by a force: "Time itself ought not to be said to exist or not to exist for some time, otherwise time would be needed for time." (*Pacidius*, p. 622). One might, of course, claim that *instants* are actual, at least instantaneously, and that a whole duration is then real if its parts are, but still no part of duration can be said to have a duration, as we have seen, and in any case Leibniz denies that instants are properly "parts of time":

> Everything which exists of time and duration, being successive, perishes continually. And how can a thing exist eternally if, to speak precisely, it never exists at all? For how can a thing exist when no part of it ever exists? Nothing of time ever exists except instants, and an instant is not even a part of time. Anyone who considers these observations will easily comprehend that time can only be an ideal thing. (To Clarke, V, #49: G. VII. 402)

Leibniz concludes this harangue with the remark that "the analogy between time and space will easily make apparent the fact that one is just as purely ideal as the other" (*ibid*). But to this we should rejoin with the obverse of his own remark in V, #74: "From duration to extension, *non valet consequentia* (the inference is invalid)". For although the above argument is sufficient to preclude the existence *in re* of any part of time or duration, there is nothing in it to derogate from the possibility of a part of *extension*'s having actuality: one could, as the Cartesians did, claim it to have parts, duration, and be sustained in existence by a force. What this demonstrates is the ***subsidiary*** nature of the above reasons for time's ideality, and their peculiarity to time.

It is clear from Leibniz's arguments with the Cartesians, however, that his principal argument for the ideality of time is the same as that for the ideality of space: it is, as already mentioned, that time and space are ***abstractions*** from concrete existents. For although he agrees with Descartes that time and space are true and real -- they "are of the nature of eternal truths" (New Essays, XIV, §26: G. V. 140), they underwrite the reality or veracity of phenomena ("Conv. b. Phil. & Arist.": G. VI. 590) -- Leibniz was adamant that abstractions, and extension in particular, were not the kind of entities that could possible by said to *exist*, or to constitute the complete natures of determinate existents. As he explained to Malebranche, being abstractions they presuppose the existence of the concrete particulars they are abstracted from:

> Extension is nothing but an Abstraction, and requires something which is extended. It needs a subject; it is something relative to this subject, as is duration. (G. VI. 584)

Secondly, because space and time are abstract entities, their parts are all indiscernible from each other: no one part of space or time can be distinguished from another without reference to some determinate, intrinsically identifiable extended or enduring subject. Thus abstactions are exempt from the Identity of Indiscernibles, which only applies to concrete existents, to actuals; in fact, it is precisely because their parts are all indiscernible that abstract entities are only ideal. But the important point here is that the parts of time are only indistinguishable when they are considered in the abstract. That is, *time is only ideal* when it is considered *apart from things, in abstracto*; but when it is individuated by some change in a thing (whether monadic or phenomenal), it is *real*, a *phenomenon bene fundatum*:

> The parts of time or place, considered in themselves, are ideal things; thus they resemble each other perfectly, like two abstract units. But it is not the same with two *concrete ones*, or with two *real times*, or two spaces *filled up*, that is to say, truly *actual*. (To Clarke, V, #27: G. VII. 395)

With regard to space, Leibniz gives a very complete account of this doctrine, explaining to Clarke (in V, #47) the difference between *concrete space*, which he later (V, #104) describes as the "order according to which situations are disposed", and *abstract space*, which is "that order of situations when they are conceived as being possible", and which "is therefore something ideal" (*ibid*: G. VII. 415). This is explained as follows: observing many things in a certain determinate *order of coexistence* -- that is, having certain fixed relations of situation to one another -- one judges something to change its place if its relations to these *fixed existents* change, whilst the order of the situations among the fixed existents themselves remains the same. Likewise A occupies the *same place* as B if its relations to these fixed existents agrees with B's. The order of situations among such a fixed set of determinate existents is a "space filled up", and is accordingly something concrete. (This is *phenomenal space*, well founded on the determinate existents it relates together.) But if one abstracts away from these existents, and considers these places and the relations among them *in abstracto*, one is now dealing with abstract entities, *entia rationis*. Similarly, space considered as "that which comprises all these places" is just abstract space; the latter "can only be ideal, containing a certain order in which the mind conceives the application of relations" (V, #47: G. VII. 401).

In the same way, Leibniz explains, the ratio or proportion between two lines L and M may be considered in three different ways:

> as the ratio of the greater L to the lesser M; as a ratio of the lesser M to the greater L; and lastly as something abstracted from both, that is to say, as the *ratio* between L and M, without considering which is the anterior or posterior, the subject or the object.... In the first way of considering them, L, the greater, is the subject; in the second, M, the lesser, is the subject of this accident, which philosophers call *relation* [*relation ou rapport*]. But which of them will be the subject in the third sense?" (To Clarke, V, #47: G. VII. 401)

Leibniz concludes that "this relation, in this third sense, is indeed *outside* the subjects; but being neither substance nor accident, it must be a purely ideal thing, the consideration of which is nevertheless useful."

This passage has, of course, been much quoted by the proponents of the relation-reducibility thesis (e.g. Russell, RuPL pp 12-13, Rescher RePL pp 95-96), who see in it a strict application of the subject-predicate logic to prove the ideality of relations. Yet as Ishiguro has argued (see Ish Ch 6, esp. pp 103-105), the *ratio* in question is for Leibniz a *relation* in all three cases; but in the third case it is considered in abstraction from the things it relates, and it is only in this case that Leibniz deems it to be ideal.

So I think it may be fairly concluded that *it is only relations considered in abstraction from determinate relata that are ideal.* As an *abstract entity,* a relation is ideal, an *ens rationis*; but inasmuch it expresses a relational property of one relatum (subject) or the other, it is a concrete attribute of the subject in question. It follows that space and time are ideal only insofar as they are orders of such abstract relations; but this by no means entails the ideality of spatial and temporal orderings of concrete existents. Leibniz is not nearly the idealist he is often portrayed to be.

Referring back to the exposition of Leibniz's theory of time I gave in Section 2, we may say that if we are given two determinate states or events, such as the birth and death of Mohammed (b and d), then the same relational fact can be expressed as b's being before d ($b <t d$), and d's being after b ($d t> b$); but that the relation of temporal precedence itself, $<t$, is something abstract, an *ens rationis.* Similarly, the time between b and d, both events having occurred in this world ($b,d \epsilon W_a$), is a "time filled up" by a continuous series of states or the phenomenal events resulting from them -- most pertinently, the states or events comprising Mohammed's life -- and is consequently a "real time", a *phenomenon bene fundatum.* (This is presumably what Leibniz meant in referring to time, like "matter taken for the mass in itself", as being "only a pure phenomenon or well founded appearance" in his letter to Arnauld of Oct 9, 1687 (G. II. 118-119).) But considered as a *quantity* of time, a stretch of time within which any possible states might have occurred, such a time is purely ideal. Again, an *instant* considered as a simultaneity class of possible monadic states, that is the class of states simultaneous with some arbitrary state, is *ideal,* a mathematical abstraction indiscernible from any other instant; and so is *time,* the family of these ideal instants successively ordered, $\{I_1, <t\}$. An instant of *real time,* on the other hand, can be identified only with respect to some determinate state or event.

This is precisely Leibniz's point in his controversy with Clarke over whether time could exist before the Creation of the Universe. The point is worth explicating a little, since Leibniz's relational philosophy of time is often taken to be simply a variant of the traditional Aristotelian relativism, according to which there is no time without actual change. This it is not: for according to Leibniz time and space are not just orders of actual existents, but of *all possible existents*:

> But Space and Time taken together comprise the order of possibilities of one whole Universe, so that these orders (Space and Time, that is) relate not only to what actually is, but also to whatever could be put in its place, just as numbers are indifferent to whatever the *res numerata* may be. And this involvement of the possible with the Existent makes a continuity which is uniform and indifferent to all division. (Reply to Bayle, 1702, G. IV. 568)

Consequently Leibniz could not follow the Scholastics in their rejection of "premundial time" on the grounds that there was no actual change before the Creation; he was obliged to find a stronger argument, one pertaining to the *possibility* of change before the Creation.

Aristotle, of course, had no such problem. Having asserted the inseparability of time and change -- a principle that Newton-Smith has appropriately dubbed "Aristotle's Principle" -- and noting that we cannot conceive of a time without conceiving an earlier time, he simply concluded that the Cosmos must be infinitely old. Here the Christian theologians could not follow him, knowing from the Bible that the world had been created only a finite time before. Faced with this dilemma, Aquinas, for instance, had argued that the time which could thus be *conceived* before the Creation must be just that, a conceptual or purely imaginary time. The alternative was to adopt the Stoics' solution of asserting the finite duration of the universe to be a mere interval of a self-existent infinite time, thus abandoning Aristotle's Principle altogether. But against this St. Augustine had already objected that it would entail what Leibniz later called a "falling back into *loose indifference*" on the part of the Divinity. It would mean, wrote Augustine, that

> God was guided by chance when he created the world in that and no earlier time...although there was no difference by which one time could be chosen in preference to another. (*The City of God*, quoted from Capek, p 180)

It would be as if God had "set the world in the very spot it occupies by accident rather than by divine reason" (*ibid.*).

Needless to say, this argument contains more than a hint of Leibniz's principle of Sufficient Reason, and Leibniz himself does not miss the opportunity to avail himself of the same argument against premundial time in his third letter to Clarke. Nevertheless, the argument from lack of sufficient reason is hardly adequate on its own to refute the idea of a premundial time, as Leibniz realized. For Newton and Clarke did not share his and Augustine's theology: for them, time and space were attributes of God, not the world, and were accordingly coextensive with him; and from their point of view, to apply the Principle of Sufficient Reason to God would be an unacceptable limitation on his power of choosing. But Leibniz's argument against the possibility of premundial time, although it makes use of this Principle, is subtly different from Augustine's, and altogether more potent. It is worth quoting in full:

> Supposing someone were to ask why God did not create everything *one year sooner*; and the same person should wish to infer from this that God has done something for which it is *not possible* that there should be a *reason*, why he did it *so* rather than *otherwise*: one would answer him that this inference would be true if time were something apart from temporal things, for it would be *impossible* that there should be *reasons* why things should have been applied to these *particular instants* rather than *others*, their succession remaining the same. But the same argument proves that *instants* apart from things are *nothing*, and that they consist only in the successive *order* of things; which order remaining the same, *one* of the two states, for instance that of the imagined anticipation, would not be any different from, and could not be discerned from, the *other* which is now. (To Clarke, III, #6: G. VII. 364)

The first half of this argument is similar to Augustine's, but not quite the same. Augustine had argued directly against the possibility that all things, events etc. in the world might have been created a certain time sooner on the grounds of lack of sufficient reason. Leibniz, in contrast, does not just dismiss this possibility, but uses it as a premise in a *reductio* argument against absolute time. He is prepared to provisionally grant the Newtonians' claim that the world might have been created sooner in order to demonstrate its absurdity. Thus he considers the state of affairs in which the date of the Creation and all subsequent events are displaced in this way as another possible state of affairs distinct from the actual one, and argues that if the temporal relations of a given event to all other events were still the same, the instant at which that event occurs would be indistinguishable in the two cases. Thus the supposedly earlier event in the hypothetical case is not in fact distinct from the corresponding event in our world, an obvious absurdity. It follows that both the instants and in fact the entire temporal orders would be identical in the two cases.

Implicitly Leibniz is here appealing to his Principle of the Identity of Indiscernibles, but in a context in which it is very difficult to deny. For if one posits temporal relations as existing among instants of absolute time, that is, independently of things and their states, and if the states of things are said to have their positions in time by occupying some such instants, then -- since in themselves instants are all identical -- there is no way of telling *which* instants they occupy; unless, of course, instants are identified by the states or events occupying them, which is Leibniz's view.

It is important to appreciate the superiority of this position over Augustine's. For if Leibniz is interpreted as simply upholding the long tradition of the Schools which denied the existence of time in the absence of change -- as Clarke interpreted him, and as many commentators still do -- then his position is easily countered. For if one chooses to *define* premundial time as "imaginary", as had Aquinas, one does "not thereby *prove*", as Clarke rightly pointed out, "that [it] is *not real*". One might equally claim that all it means is "that we are wholly ignorant of *what kinds of things [take place]* in that [time]".[25] In other words, one cannot prove that the possibility of the world's being created sooner is not a *real* possibility by stipulation, since it is equally possible to stipulate the contrary, as did Gassendi, following the lead of the Stoics.[26]

For Leibniz, on the other hand, the impossibility of premundial time is not a consequence of the a priori identification of time with the duration of the actual universe -- indeed, as we have seen, time relates equally to all *possible* changes in the universe. However, in the absence of any possible change or event which might individuate it, an instant of this time of possible change is simply *ideal*, an abstraction in the minds of men. Thus Leibniz, by means of his argument from indiscernibility, does not *assume* but, as he himself says, "*demonstrate[s]* that time without things is nothing more than a mere ideal possibility" (To Clarke, V, #55: G. VII. 404). For on his relational view, the possibility of a given event occurring at a different time is represented by its bearing different temporal relations to the other events in the temporal order. (This is precisely analogous to Leibniz's interpretation of change of place of a body as a change of its location within the order of situations.) But the Creation of the Universe is not such an event, since it cannot change place within the temporal order: no possible event could occur before the beginning of the world because it would then *be* the beginning of the world ("The beginning, *whenever* it was, is always the same thing" (IV, #15: G. VII. 373-374). Therefore

> [I]f anyone should say that this same world which has actually been created could have been created *sooner*, without any other change, he would be saying *nothing intelligible*; for there would be no mark or difference by which it would be *possible* to know that it had been created *sooner*. (To Clarke, V, #55, G. VII. 405)

Thus we see that no elaborate reconstruction of Leibniz's position in terms of possible world semantics or counterfactuals is necessary for an understanding of it. Indeed, the very point of the above argument is that a world in which the mutual temporal relations among all the states or events were the same, but in which any of the states or events occurred sooner, is *not* a possible one, since there is no criterion of individuation by which one could discern different instants within a given temporal order save by their correspondence with states or events.

The power of Leibniz's position, finally, shows up well in a comparison with the views expressed by Newton in his early manuscript *De Gravitatione*,[27] which has only recently been published. There Newton seems to anticipate Leibniz in his conception of time as an order, and also in his claim that instants are only individuated by their place in this order:

> For...the parts of duration derive their individuality from their order, so that (for instance) if yesterday could change places with today and become the later of the two, it would lose its individuality and be no longer yesterday but today....By their mutual order and position alone are the parts of duration and space understood to be just what they are in fact; nor have they any other principle of individuation besides that order and those positions, which therefore they cannot change. (*De Gravitatione*, translation from Stein, p. 194)

These similarities have prompted a recent trend among commentators to play down the differences between Newton's and Leibniz's philosophies of time. But a closer inspection shows that the supposed similarities are not as great as they first appear. For although Newton agrees with Leibniz that "All things are placed in time as to order of succession" (*Principia*), for him this does not function as a definition of time, as it does for

Leibniz. The "parts of duration", despite obtaining their individuality only through their positions in the order, are not to be equated with the temporal locations of any *thing*, actual or possible: for duration is a property (or mode of existence) of God, who exists independently of things. Accordingly there is a great deal of difference in the nature of the order of the parts of time as conceived by the two thinkers: for whereas Leibniz conceives the order of succession as obtaining only among determinate states, and derivatively among the (real) instants individuated by these states, Newton conceives the order as obtaining among the parts of time considered *in abstraction from any things or states* which might individuate them: they derive their individuality from their positions in this order alone. Thus whereas for Leibniz the parts of time, considered in themselves, are completely indiscernible and similar to each other, and therefore *homogeneous* -- each in fact being identical with the ideal temporal continuum -- for Newton these parts are intrinsically distinguishable by their place in an order which exists absolutely (presumably in the mind of God), and are therefore *inhomogeneous*. It is this very inhomogeneity or individuality of their parts prior to their being filled with things or events, which makes for the immutability and immovability of absolute time and space in Newton's view:

> As the order of the parts of time is immutable, so also is the order of the parts of space. Suppose those parts to be moved out of their places, and they will be moved (if the expression be allowed) out of themselves. For times and places are, as it were, the places as well of themselves as of all other things.[28]

In modern physics the homogeneity of time is usually represented by the requirement of the invariance of all laws and equations under time translation. That is, it is acknowledged that in a homogeneous time the translation of all dates and times of all events and states say, backwards one year, can make no real difference to the world. Leibniz's anticipation of this invariance requirement, and indeed his implicit anticipation of the connection between this characterisitc of time and the need for a constant measure of total activity, are, I believe, to be reckoned among his greatest achievements.[29]

Summarizing the conclusions of this section, we may say first that it appears that Russell was wrong in his belief that Leibniz held all relations to be reducible to monadic properties: temporal relations between monadic states could not be reducible to unary temporal properties of states without making nonsense of the major part of Leibniz's writing on time, where the temporal order of every monadic series is founded in the order of reasons or pre-established harmony. Secondly I argued that space and time are ideal for Leibniz not because they are relations and thus inapplicable to actuals, but because, considered in themselves, they are abstractions from existing things rather than things existing in their own right (as Descartes and Newton supposed). Furthermore, no part of time can be individuated save by reference to a state or event occurring at that time: considered in the abstract each part of time is perfectly similar to every other and to the whole. In contrast to this "abstract time", Leibniz also considers a "concrete time", by which he means a time whose parts are individuated by states or events occurring at them, that is a temporal ordering of existents. Thus it is only elements of concrete time that can be distinguished from each other; considered in itself, time is perfectly homogeneous.[30] Lastly, since times can only be individuated by the states and events they relate, the latter must be invariant under a translation of the origin of time. Thus Leibniz's theory has physical content: it would be refuted if physical laws were not invariant under time translation.

4. Time, Phenomenal Change, and Cause

In section 2 I interpreted Leibniz's theory of time to be based on relations among monadic states, as opposed to the usual view that Leibniz's time consists solely in relations among phenomena; and in the last section I argued that such monadic temporal relations are not precluded by the ideality of time, as is often supposed, since it is only when it is considered in abstraction from things that time is ideal. I also implicitly assumed there that the same time concept could be used to order phenomenal events. But we still have no inkling as to why Leibniz should have founded time on monadic series of states rather than founding it directly on observable physical events, nor have we seen how it is that a monadically founded time can justifiably be applied to phenomena. It is these issues that I want to address in this section.

As I have presented it above, Leibniz's theory of time is essentially *metaphysical*: it involves the postulation of entities which cannot be directly observed (the monads and their states) and founds temporal succession on an ideal relation of reason-inclusion or ground-containment, a relation which cannot in general be discerned to obtain because it requires an infinite analysis. How, then, would Leibniz be justified in using the same time concept to refer to *phenomenal* change and succession, as we have assumed above? Would he not have to construct a separate time concept, based on the ordering of phenomena? And do not his references to phenomenal time and to time as "the order of successive phenomena" demonstrate that he did indeed have such a conception of time, distinct from the inter-monadic time described above?

I do not think that Leibniz either had or needed to have such a distinct time concept. For, given the dependence of time upon change, to assume that Leibniz had a concept of phenomenal time *distinct* from his monadic time is tantamount to severing the connection between phenomenal change and monadic change. (Some interpreters, of course, see in this a difficulty for the very concept of monadic change: according to them, since it cannot occur in phenomenal time, which is ideal because relational, monadic change cannot occur at all.) But Leibniz was quite adamant that monadic change is prior to phenomenal changes; thus his justification for applying the same time concept to phenomenal changes as to monadic changes of state is that the former are *results* of, and are grounded in, the latter. All things change, and composite things (phenomena) change because of the changes of the simple things from which they result. Leibniz is explicit on this point:

> You doubt, distinguished Sir, whether a single, simple thing would be subject to changes. But since only simple things are true things, the rest being only Beings by aggregation and thus phenomena, and existing, as Democritus put it, νομω not φυσει, it is obvious that unless there is change in the simple things, there will be no change in things at all. (To de Volder, 20/6/1703: G. II. 252)

The precise way in which phenomena are supposed to result from monads and their changes of state is, however, one of the more obscure points of Leibniz's system. In his earlier work, where he conceived of monads as located at points in a real space, Leibniz conceived of phenomenal bodies as infinite aggregates of monads. This conception persists in his later work, though alongside of it we find accounts which suggest that phenomena are rather aggregates of perceptions of states. This tends to give the impression that Leibniz moved from an outright realism in his youth, with bodies *compounded* out of atoms of substance, to a mature phenomenalism where everything is a perception and nothing is directly perceived, a phenomenalism tempered (albeit inconsistently) by his references to monads as "contained in" the bodies which result from them. But this impression should be resisted. As I have argued elsewhere[31], after 1670 Leibniz consistently rejected the idea that bodies could be *composed* out of simple substances on logical and mathematical grounds.[32] The basis of this rejection was his denial of the possibility of an *infinite collection*: an infinity of terms, according to him, could only be understood as a whole in a *distributive sense*, as a set of terms related together by some defining property or law. In the case of an infinity of monads, this mutual relation is supplied only in perception, so that an infinite aggregate of them constitutes a *perceived whole*, rather than a true one. In his mature philosophy, on the other hand, there exist wholes whose reality does not derive merely from their being perceived, namely the monads, whose states are connected together by the law of their series, each following the other according to the preestablished order of reasons. Thus Leibniz's position is neither straightforwardly realist nor phenomenalist, but somewhere between.

This doesn't take us very far, of course. For it is still not very clear how phenomenal bodies may be said to *result* from an aggregate of monads. I shall not attempt to resolve this problem here, although the following visual analogy may go some way towards relieving the obscurity of the idea.

Imagine a table standing in a room. The room would look different from every point on the surface of the table, but the differences in perspective would be smaller the closer together the points were from which the room was viewed. Moreover, there would be a correspondence between the perspectives, or the visual perceptions of an imaginary eye ranging across the table top, and the situations of the imaginary eye in the table top.[33] Thus any changes taking place in the room would be represented in the different perceptions of the room by the imaginary eyes at different times, there being a kind of harmony or agreement among the perceptions of all the eyes at some given time.

Similarly for Leibniz everything which exists in the universe is represented (more or less confusedly) in any given monad from its own point of view, this representation being what Leibniz calls a *state* or *perception* of the monad. As we have already seen, it is the harmony among the states of all the monads in the world which constitute a given real time; and it is the connection among all the states of the same monad through time (by virtue of which each follows from the others according to the law of the series) which constitutes the monad as the same perceiving being.[34] But just as there is an agreement between the sum total of perspectives perceived by the imaginary eyes in the table top at different moments of the table's existence, so too there is an agreement among the perceptions of all the constituent monads of any well founded phenomenon at different moments of the phenomenon's duration; and likewise, because of their reciprocal nature of the universal harmony (all coexistents represent each other), there will also be agreement among the perceptions of the phenomenon by any other monad in the world at different moments of the phenomenon's existence, just as the "same" table would be seen by imaginary eyes in the walls of the room at different times. Now it is these agreements among the perceptions of (or by) all the monads in a given aggregate which consititute this aggregate as a lasting phenomenon, and the changes in such an order of perceptions which constitute a change in that phenomenon. In this way, not only phenomenal durations, but phenomenal changes too, have their foundation in the contingent orderings of qualitative similarity among the perceptions or states of a given aggregate, and the changes in such orderings. Thus it is that phenomenal duration and change are *well-founded* in changes of monadic states.

This may help to explain how Leibniz could refer to time sometimes as an order of successive *states*, sometimes as an order of successive *phenomena*, (and more generally as an order of successive *things* (*successivorum*)), without embarrassment. But it still doesn't explain *why* he didn't give an account of time *purely in terms of phenomenal change and the order of cause and effect*, why he saw it as necessary to found time and change at the monadic level. A full answer to this problem depends on an understanding of his solution to the paradoxes of the continuity of time, which I have explored elsewhere. But an investigation here of the difficulties of a purely phenomenal time (in Leibniz's sense) will help to clarify his reasons for introducing monads and their changes.

In section 2 I mentioned that the standard interpretation of the relational theory of time implicit in the *Initia Rerum Mathematicarum Metaphysica* is in terms of relations of cause and effect among phenomena. In this respect van Fraassen's estimation of Leibniz's theory as "the first attempt at a *causal theory of time*" is typical. Commenting on the passages in question van Fraassen writes:

> In other words, according to Leibniz the various

> circumstances or states of affairs are related to each other as cause to effect, and by definition, the cause is earlier. (vanF, p. 38)

Similarly Winnie writes that Leibniz "seems to have been the first to conjecture that the structure of time and space might be reducible to causality" (Winnie, p 136). As a result, Winnie gives an account of Leibniz's theory which is formally almost the exact counterpart of that given above (his is the original), but which is interpreted in terms of *events* and *causal connections* (p 138). His version of Leibniz's postulate of the connection of all things, for instance, (Axiom 4 above, which was needed for the transitivity of simultaneity), runs as follows:

> *Postulate II (Leibniz Postulate)*: If e_1 and e_2 are simultaneous and e_1 is causally connected with e_3, then e_2 is also causally connected with e_3. (Winnie, p 138)

Now it cannot be denied that the principle of "the connection of all things" is understood by Leibniz to apply to material phenomena or compounds as well as to monads or simple substances: since all causal action is by contact and the universe is a plenum, every causal action will have some effect on all subsequent phenomena, as Leibniz states explicitly in the *Monadology*:

> And composite substances symbolize with simple ones in this respect. For since the universe is a plenum, with the result that all matter is connected together, and since all motion in the plenum has some effect on distant bodies in proportion to their distance, so that each body is not only affected by those touching it and in some way feels the effect of everything that happens to them, but also by their mediation feels the effect of those bodies which are touching the ones with which it is in immediate contact: so it follows that this communication extends to any distance whatever. (*Monadology*, #61: G. VI. 617)

Nevertheless it can be stated with some confidence that Leibniz would never have tried to *reduce* temporal relations to causal ones, or to *define* time in terms of causality. In the first place, the whole point of Leibniz's appeal to metaphysical principles is for the explanation and grounding of matters which he believed could not be explained and grounded purely in terms of phenomena -- matters such as the problem of the self-identity of a substance through time, and the continuity of motion. Since, as we shall see, one of the chief motivations for Leibniz's introduction of monads undergoing a continuous succession of states in time was a perceived inability to explain causal action on the phenomenal level, it would make nonsense of the whole strategy of recourse to metaphysical principles if this succession of states in time then had to be reduced to (or defined in terms of) causal actions among phenomenal bodies.

Secondly, and consequently, on Leibniz's scheme the term "causal action" must in any case be understood metaphorically. To assert that one created substance "exerts physical action" on all others is simply to say that "if a change occurs in one, some corresponding change occurs in all others" (*Primae Veritates*, Cout. OF p 521, (Loem, p 269)). But

> What we call causes are, in Metaphysical rigour, merely concomitant requisites....For to say nothing of the fact that it cannot be explained how anything can pass from one thing into the substance of another, it has already been shown that from the notion of every single thing all its future states already follow. (*ibid.*)

The roots of this view of causality are already in place in Leibniz's thought as early as 1676, before he had introduced his simple substances into the explanation of motion. In his brilliant analysis of continuous motion in the dialogue *Pacidius Philalethi* (Cout. OF pp 594-627) of October 1676, Leibniz had already discovered that purely material bodies "do not act while they are in motion". His reasoning was as follows. The last moment of one state of a thing (e.g. the last moment of life) cannot be identical with the first moment of its next state (e.g. the first moment of death), since then the same thing would be in two contradictory states at the same time. Consequently these two moments must be merely *contiguous*, as Aristotle had argued. But it follows from this that no matter how many different states a given continuous change or motion is divided into,

> there is no moment of change common to each state, and so neither is there any state of change, but only an aggregate of two states, old and new; and so there is no state of action in a body, no moment can be assigned at which it acts... (*Pacidius*, Cout. OF p. 623)

As an example of this, Leibniz considers a body E moving across the boundary between two contiguous spheres, AB and DC. At one moment

E is the sphere AB, and at the next moment it is in DC. But at neither moment was there a state of action or transference of the moving body from one sphere into the other. Leibniz concludes:

> Thus action in a body cannot be understood except through a certain aversion. If you really cut to the quick and inspect every single moment, there is none. Hence it follows that proper and momentaneous Actions belong to those things which by acting do not change. And therefore the action...by which the moving body which was in one sphere at one moment is caused to be in the other contiguous one at the following moment, does not belong to the body being transferred, E. (*Pacidius*, Cout. OF p 623)

Here then is the origin of Leibniz's view that the physical action of one body on another can only be understood in a metaphorical sense. For on the level of ***bodies*** -- that is, of phenomena -- all one ever finds is change in the sense of ***difference*** between one state and the next, but never a ***state of action***, a cause or principle by which motion is transferred. So where is this cause to be found? When Leibniz wrote the *Pacidius* just after the end of his four years in Paris, he was still under the influence of the Cartesian Occasionalists, and had no hesitation in adopting their solution: the continuous, direct action of *God* was necessary to sustain motion.

> Therefore that by which the body is moved and is transferred is not the same body, but a superior cause which by acting is not changed, which we call God. Whence it is clear that a body is not even able to perform continuous motion of its own accord; but needs the continuous impulse of God, who, however, acts constantly and in accordance with certain laws by virtue of his supreme wisdom. (*Pacidius*, Cout. OF pp 623/4)

The conclusion that the one superior cause must be God is one that Leibniz -- and, of course, a good many other philosophers -- had reached before. This is not nearly so interesting as his previous way of wording the conclusion to this argument, namely that "proper and momentaneous Actions belong to those things" -- note the plural -- "which by acting do not change". For this description, in the context of his later philosophy, would serve as a tidy definition of his simple substances or Monads. Certainly, it is not too speculative to see here one of the principal reasons for their introduction. Indeed one of the most fascinating aspects of the *Pacidius* is the fact that it goes against the grain of much of Leibniz's earlier philosophy. The Occasionalist philosophy he espouses here, which makes God the sole source of activity, and matter completely inert, contrasts sharply with Leibniz's earlier (and later!) attempts to uphold the Aristotelian conception of substance as something which *acts.*[35] The tension between this conception of substance and the absence of activity in phenomenal bodies is finally resolved in Leibniz's new concept of the action of a substance as the inherent tendency of each of its states to spontaneously pass into the next, which he later termed *appetition*, the successive states being determined by the law or notion of the substance which remains unchanged while the substance acts.

However, this mature conception of substantial action, and the concomitant move from the one superior cause of Occasionalism to the positing of a cause of motion in every single part of matter, had to await the development of Leibniz's dynamics. It was only after he had distinguished the *motive force* of bodies from their instantaneous actions, and had identified the latter, the *conatus*, as instantaneous elements of motive *force* rather than of *motion*, that Leibniz was prepared to fully relinquish Occasionalism.

There is one further point to be made in connection with the *Pacidius* concerning Leibniz's philosophy of causation and its effect on his theory of time. This concerns the fact that in this dialogue, and indeed in all his prior work too, Leibniz presupposes the *reality of space*. Now the point is that if space is real, and motion is change of location in space, then it follows that motion too is something real and *absolute*: there can never be any ambiguity over which of two bodies in relative motion is really moving. However once one accepts the *relativity of motion*, as Leibniz was shortly to do -- a relativity understood in the sense of the "Equipollence of Hypotheses", the complete indifference of the resultant phenomena to which of them is hypothesized to be at rest -- then the idea of determinate locations in a substantial space loses all its physical significance. It is not clear precisely when (between 1676 and 1686) Leibniz abandoned his assumption of the reality of space, nor whether it was the arguments we reviewed in the last section concerning the abstract nature of space, or his appreciation of the significance of the relativity of motion, or perhaps some admixture of criticisms, which first led him to abandon it. But whatever the reasons, by the time he had established the key concepts of his dynamics in 1686, the realization that the relativity of motion made the question of what was the true subject of a given relative motion problematic was in the forefront of his thoughts. In fact this point was one of the two criticisms he jotted in the margin of the *Pacidius* that year:

> There still remains to be treated, first, the subject of motion, so that we may tell which of two bodies changing their mutual situation motion should be ascribed to; and secondly, what is truly the cause of motion, or motive force. (*Pacidius*, Cout. OF p 594)

The relevance of this is as follows. In the *Pacidius*, if a moving body collides with another at rest, one can say that the former is the cause of motion of the latter -- "the prior harbours the cause, the posterior expresses it", p 624) -- provided only that one understands 'cause' in a metaphorical sense. With this proviso, then, it would in principle be possible to set up a "causal" theory of time at the phenomenal level. (In strict metaphysical rigour, of course, there is no necessary connection between body A's being in motion immediately prior to the collision and body B's being in motion immediately after, any more than there is between two consecutive states of motion of a single body. For "the following state does not follow from the preceding one with necessity", and the cause of the two states can only be understood as "a certain permanent substance which both destroyed the first one and produced the new one" (p 624).)

But once the relativity of motion is accepted -- that is, the equipollence of hypotheses -- the possibility of such a causal theory on the phenomenal level completely disappears. For now

> It is a consequence that not even an Angel could discern, in mathematical rigour, which of several such bodies [in mutual motion] is at rest, and is the centre of motion of the others. (*Phoranomus*, [1688], Cout. OF p 590)

Thus so long as physical causality is understood in terms of one moving body colliding with another, the relativity of their motion makes it impossible to discern, on the phenomenal level, which is really moving, and thus which is really the cause. Thus it is that the possibility of a purely phenomenal theory of time is, for Leibniz, finally foreclosed.

But this argument from the relativity of motion appears to be a secondary consideration in Leibniz's abandonment of a theory of time based on changes among phenomena; the primary reason for it, as we have seen above, is his conviction that it is impossible to explain activity and its continuity while remaining on a purely phenomenal level.

5. Conclusion

The interpretation of Leibniz's theory of time which I have argued for here differs markedly from previous ones in construing time principally as a structure of relations among ***monadic states***, and only derivatively as a structure of relations among phenomena. This construal has the immediate advantage of making explicit Leibniz's founding of time in the preestablished order of reasons: the temporal order of states of a given monad corresponds to their ordering according to the relation of reason-inclusion, and the compatibility of simultaneous states is underwritten by the preestablished harmony of any given possible world. I argued in section 1 that this interpretation absolves Leibniz from the criticism that his relational theory presupposes an internal absolute or "intra-monadic" time for each monad -- the second set of difficulties noted in the introduction. I also argued that the correspondence between temporal and rational orders does not amount to a reduction of time to atemporal relations, since Leibniz posits in addition a principle of becoming for monadic states.

With regard to the difficulties concerning Leibniz's advocation of the *ideality of time*, I argued in section 2 that this ideality is not a consequence of any doctrine of the reducibility of relations to unary monadic predicates, as Russell and others have mistakenly supposed, but results rather from the ideal nature of relations considered in abstraction from things; the ideality of time pertains only to time as an *abstract entity*, and does not preclude the states of actual substances, or the events resulting from them, from occurring in real temporal relations to each other. Thus time considered in abstraction from things, *abstract time*, is distinguished by Leibniz from temporal orderings of actual states or events, which he calls *concrete times*. Whereas parts and instants of concrete time are distinguishable by reference to the states and events occurring at them, the parts and instants of abstract time are perfectly similar and are in themselves indiscernible. Therefore time in itself, Leibniz concludes against Newton, is perfectly homogeneous, and cannot exist *in re*.

Finally I explored Leibniz's reasons for founding time at the monadic level rather than construing it directly as a relational structure of phenomena. I concluded that as early as 1676 Leibniz's analysis of phenomenal change had convinced him that the cause or principle of change had to be sought on a supraphenomenal level, namely in "those things which by acting do not change", thus precluding a theory of time founded on phenomenal changes alone. Since phenomenal changes result from aggregations of monadic states, however, they are ordered by the same time concept as the states they are founded on. This resolves the remaining difficulties concerning the relationship between phenomenal and monadic times noted in the introduction.

Appendix: A Formal Exposition of the Theory

Suppose M is the set of all possible monads m, i.e. $M = \{m\}$, and let us denote the series of states of some monad m by the indexed set S_m. Now we know that for any such series of monadic states S_m, all the states subsequent to some given state in the series are dictated by the law of the series: the given state *involves the reason for*, or *contains the ground for*, all the subsequent states. Let us denote this relation of ground containment holding between any two states a and b ($a,b \in S_m$) -- a contains the ground for b -- by aGb. Then two properties of G, its transitivity and its asymmetry[36], may be postulated without further ado:

Axiom 1 (Transitivity of G)

$$(\forall a,b,c \in S_m)\ aGb \ \&\ bGc \rightarrow aGc$$

Axiom 2 (Asymmetry of G)

$$(\forall a,b \in S_m)\ aGb \rightarrow \sim bGa$$

From these two axioms alone it immediately follows that no state contains its own ground:

Theorem 1 (Irreflexivity of G)

$$(\forall a \in S_m) \sim aGa$$

We may now capture Leibniz's definition of simultaneity by defining simultaneous states as those which are not incompatible. But first we need to make explicit the fact -- implicit in Leibniz's treatment in the *Initia Rerum* -- that all states of the same monad are mutually exclusive:

Axiom 3 (Incompatibility of different states of the same monad)

$$(\forall a,b \in S_m)\ aGb \vee bGa = aIncb$$

This serves to introduce a new binary relation, that of *incompatibility* (*Inc*), holding between pairs of monadic states. From Axiom 3 and Theorem 1 it is easy to prove the *irreflexivity* (*Theorem 2*) and *symmetry* (*Theorem 3*) of *Inc*. Assuming now that we can somehow extend this conception of incompatibility to the set of states of all the monads in a given universe or world W, we shall then be able to render Leibniz's definition of simultaneity (*Sim*) as follows:

Definition 1 (Simultaneity as non-incompatibility)

$$(\forall a,b \in W)\ aSimb = \sim aIncb$$

The set W will be defined in due course. Now

Theorem 4 (Simultaneity as mutual lack of ground)

$$(\forall a,b \in W)\ aSimb = \sim aGb \ \&\ \sim bGa$$

follows immediately from Axiom 3 and Definition 1: it could, of course, have been postulated instead of them as a simpler definition of simultaneity obviating the need for the introduction of *Inc* as a separate relation.

We are now in a position to give a relational definition of an *instant*, and of *temporal order*. An instant will be the set of all monadic states simultaneous with a given state; to happen at an instant will be to be a member of this set, and thus to be simultaneous with a given state:

Definition 2 (Instant I_1)

$$(\forall I_1 \in W)\ I_1 \text{ is an instant} = (\exists a \in W)\ I_1 = \{b|aSimb\}$$

Definition 3 (Temporal precedence $<t$)

$$3(\text{i})\quad (\forall I_1,I_2 \in \{I_1\})\ I_1 \leqslant tI_2 = (\exists a \in I_1, \exists b \in I_2)\ (aGb)$$

$$3(\text{ii})\ (\forall a,b \in W)\ a \leqslant tb = (\exists I_1,I_2 \in \{I_1\})\ (a \in I_1\ \&\ b \in I_2\ \&\ I_1 \leqslant tI_2)$$

Definition 3(i) gives us Leibniz's rendition of time as "the order of existence of the non-simultaneous" (*Initia Rerum*, GM. VII. 18), or as he also called it, "the order of inconsistent possibilities" (To de Volder, 20/6/1703: G. II. 253). It is the family of instants $\{I_1\}$ -- that is, the quotient of the set of all the states in the world, W, by the relation of simultaneity *Sim* -- ordered by temporal precedence. Definition 3(ii) extends the definition of temporal precedence to ***states***; this allows us to deduce that a*G*b entails $a <_t b$: if a involves the reason for b, a is the temporal antecedent, b the consequent.

But all of this presupposes that instants defined as above will constitute ***distinct*** classes of states: formally, that simultaneity is an ***equivalence relation.*** Now, from Theorems 2,3, and 4 the reflexivity and symmetry of *Sim* follow easily:

Theorem 5 (Reflexivity of *Sim*)

$$(\forall a \in W) \; a\mathit{Sim}a$$

Theorem 6 (Symmetry of *Sim*)

$$(\forall a,b \in W) \; a\mathit{Sim}b \rightarrow b\mathit{Sim}a$$

For *Sim* to be an equivalence relation, however, it must be not only symmetric and reflexive, but ***transitive***, a property which cannot be deduced from the axioms and definitions given so far. What we need, in fact, is some way of extending the concept of incompatibility across states of ***different*** monads. As explained in the text, Leibniz recognized this, and supplied the need with his hypothesis of "the connection of all things", which posits a harmony among all the states of any given world W:

Axiom 4 (Connection of all things)[37]

$$(\forall a,b,c \in W) \; a\mathit{Sim}b \;\&\; bGc \rightarrow aGc$$

In conjunction with Theorem 4 this yields the required transitivity of *Sim*[38]:

Theorem 7 (Transitivity of *Sim*)

$$(\forall a,b,c \in W) \; a\mathit{Sim}b \;\&\; b\mathit{Sim}c \rightarrow a\mathit{Sim}c$$

With simultaneity thus established as an equivalence relation, Theorem 4 now asserts the ***(weak) connectivity*** of the ordering *G* on the set W in the following sense: if neither of two states involves the reason for the other, then they occur at (are members of) the same instant. Time itself will now consist in a *total ordering* of instants by the relation of temporal precedence $<t$, thus confirming Leibniz's contention that every state will be either earlier than, later than, or simultaneous with any other.

Leibniz's category of *compossibility* may now be captured as follows. Suppose U is the set of states of all possible monads, $U = \bigcup_{m \epsilon M} S_m$. Now any two monadic series S_I and S_J will be *compossible* iff for any state of S_I there is a unique compatible state of S_J:

Definition 4 (Compossibility of Monadic Series of States)

$$(\forall S_I, S_J \epsilon U)\, S_I \mathit{comp} S_J = (\forall a_I \epsilon\ S_I)[(\exists a_J \epsilon S_J)\,(\sim a_J \mathit{Inc} a_I)\ \&$$

$$(\forall b_J \epsilon S_J)\,(\sim b_J\ \mathit{Inc}\ a_I \rightarrow b_J = a_J)]$$

Now clearly this relation will induce a partition of the total set U, the union of all possible monadic series, into a family of distinct sets of compossible monadic series. Such a distinct set of possible monadic series is exactly what Leibniz means by a *possible world*. Thus we have

Definition 5 (Possible World W_p)

$$W_p = \{S_m | Sm \mathit{comp} S_p\}$$

Corollary: $\{W_p\} = U/\mathit{comp}$

It is on W, we recall, that time is defined.

Actually the above exposition requires some amendment, since the definition of possible worlds presupposes that the relation of compossibility is an equivalence relation on U. But this will be so only if the compatibility ($\sim$*Inc*) of states is transitive across states of monads belonging to *different possible worlds*, which is an additional implicit assumption. This is equivalent to the assumption that the entire universe of possibles can indeed be assumed to be a union of self-consistent possible worlds. But even without assuming this, it remains true that time, on Leibniz's definition, is the order of succession in any possible universe.

Department of Applied Mathematics
University of Western Ontario

Notes

For the texts most often cited, I have used the following abbreviations:

Capek	*Concepts of Space and Time,* ed. Milic Capek, Dordrecht, Boston: D. Reidel, 1976.
Cout.OF	*Opuscles et fragments inedits de Leibniz,* ed. Louis Couturat, Paris: Felix Alcan, 1903.
G.V.p.	*Die philosophischen Schriften von G.W. Leibniz,* ed. C.I. Gerhardt, 7 vols., Berlin: Weidmann, 1875-1890, volume V, page p.
Ish	*Leibniz's Philosophy of Logic and Language,* Hidé Ishiguro, Ithica N.Y.: Cornell U. P., 1972.
GM.V.p.	*Leibnizens Mathematische Schriften,* ed. C.I. Gerhardt, 7 vols, Berlin and Halle, Weidmann, 1849-1855, volume V, page p.
LCIE	*Leibniz: Critical and Interpretative Essays,* ed Michael Hooker, Minneapolis: University of Minnesota Press, 1982.
Loem	*Gottfried Wilhelm Leibniz: Philosophical Papers and Letters,* ed. and trans. Leroy E. Loemker, 2nd edition, Dordrecht: D. Reidel, 1970.
McGuire	"'Labyrinthus Continui': Leibniz on Substance, Activity and Matter", J.E. McGuire, pp 290-326 in *Motion and Time, Space and Matter,* ed. K. Machamer and R.G. Turnbull, Columbus: Ohio State University Press, 1976.
RePL	*The Philosophy of Leibniz,* Nicholas Rescher, Englewood Cliffs, N.J.: Prentice-Hall, 1967.
RuPL	*A Critical Examination of the Philosophy of Leibniz,* Bertrand Russell, London: Allen & Unwin, 1900, 2nd. edition 1939.

vanF *An Introduction to the Philosophy of Time and Space*, Bas van Fraassen, New York: Random House, 1970.

Winnie "The Causal Theory of Space-Time", John A. Winnie, pp 13-205 in *Foundations of Space-Time Theories*, ed. John Earman, Clark Glymour and John Stachel, Minneapolis: University of Minnesota Press, 1977.

All translations from the Latin and French are my own, except those passages for which I was unable to obtain the original language source, where I have indicated the translation used. Wherever possible I have given the date of the passage cited, using the day/month/year format (so that 11/10/1705 designates the 11th October, 1705).

With regard to the paper itself, I am grateful to Don Ross and Kathleen Okruhlik for their helpful suggestions for improving the final draft.

1. See RuPL, Louis Couturat's *La logique de Leibniz*, Paris: Alcan, 1901, RePL and Ish.
2. Thus in the *Monadology* Leibniz writes: "The passing state tiewhich enfolds and represents a multitude in unity or in the simple substance is merely what is called *perception*" (Monadology, #14, [1714]: G. VI. 608) Elsewhere, however, he implies that the state of a monad contains a multitude of minute perceptions at any time, as in his reply of July 1698 to Pierre Bayle's criticisms (G.IV. 517-524) and his letter to Jean Bernoulli (21/2/1699: GM. III. 574-575). For further discussion of Leibniz's theory of perception, see Mark Kulstad, "Some difficulties in Leibniz's Definition of Perception", pp 65-78 in LCIE.
3. Rescher's interpretation is propounded in his succinct introduction to Leibniz's philosophy, RePL, and McGuire's in his recent penetrating critique of Leibniz's conceptions of substance, activity and matter, "Labyrinthus Continui" (McGuire).
4. "Monads and the Labyrinth", as yet unpublished. See also the paper cited in footnote 15 below.
5. See Russell's *Our Knowledge of the External World*, London: Allen & Unwin, 1914, 1952, and his "On the Experience of

Time", *The Monist*, 25, 1915, pp 212-233, and "On Order in Time", *Proceedings of the Cambridge Philosophical Society 32*, May 1936, pp 216-228. It is interesting to note that Russell retracts none of his criticisms of Leibniz's relationalism in the preface to the second edition of RuPL in 1937, and never acknowledges indebtedness to him on this score in his papers on temporal order. Russell's criticisms of Leibniz on dynamical grounds, on the other hand, contain the same confusion that exists in Leibniz's writings, namely that between the *relational* nature of space and time and the *relativity* of motion to a particular frame of reference. It has been one of the great achievements of 20th century philosophy of science to demonstrate the independence of these two issues. But although Russell, like Leibniz, conflated 'relational' with 'relative' he almost put his finger on a fatal flaw of Leibniz's theory: its inability to sustain a definition of sameness of place through time, or *affine connection.* For details of these points, see Howard Stein's "Newtonian Space-Time", *The Texas Quarterly*, Autumn 1967, pp 174-200.

6. Russell argues: "The definition of *one* state of a substance seems impossible without time. A state is not simple, on the contrary it is infinitely complex. It contains traces of all past states, and is big with all future states. It is further a reflection of all simultaneous states of other substances. Thus no way remains of defining one state, except as the state at one time." (RuPL, p 52). Russell assumes here what he is supposed to be proving: that simultaneity cannot be defined except by presupposing an absolute time. Leibniz, on the contrary, is guilty of no such circular reasoning, since he defines simultaneity in terms of *compatibility* of states.
7. Although it is clear that Leibniz always regards the relation G as asymmetric, it is hard to see how this is justified by the appeal to the law of the series: for this law would determine not only all the states which come after any given state in the series, but equally all those which precede it too. The lack of explanation why G should be regarded as asymmetric must therefore be considered as a serious deficiency in Leibniz's rationalistic foundation for time.

8. Compare this postulate with that of John Winnie quoted below (Winnie p 138). This excellent article prompted much of my thought on Leibniz's theory of time, and served as the prototype for this formalization.
9. The proof is as follows: suppose a*Sim*b and b*Sim*c, yet ~a*Sim*c. Then ~a*Sim*c and Theorem 4 together yield ~(~a*G*c & ~c*G*a), that is a*G*c v c*G*a by De Morgan's law. Suppose a*G*c; by Theorem 6 a*Sim*b gives us b*Sim*a, and by Axiom 4 b*Sim*a and a*G*c give us b*G*c. But this contradicts b*Sim*c (by Theorem 4), therefore not. So suppose c*G*a; by Axiom 4, b*Sim*c and c*G*a yield b*G*a. But by Theorem 4 this contradicts a*Sim*b, therefore not. a*Sim*c now follows by reductio ad absurdum, and Theorem 7 by a conditional proof.
10. Thus Barrow (who is in a sense the first of the Newtonians): "But as Magnitudes themselves are absolute *Quantums* independent on all Kinds of Measure, tho' indeed we cannot tell what their Quantity is, unless we measure them; so Time is likewise a *Quantum* in itself, tho' in order to find the Quantity of it, we are oblig'd to call Motion to our Assistance...." (*Geometrical Lectures*, Capek, p. 204.) Newton's conception is more complicated, in that Absolute Time is distinguished from the time whose quantity is measured by motion, its quantity being determined mathematically by the correction for the inequality of natural days, as I have argued in an unpublished paper, "The Theory of Fluxions and Newton's Philosophy of Time" (1981). But his criticism of Leibniz, as expressed through Clarke's pen, is still that "Space and Time are Quantities; which Situation and Order are not" (Clarke's Third Reply to Leibniz, #4: G.VII.369).
11. e.g. by van Fraassen in vanF, p 71.
12. See footnote 4 above.
13. G.II.515 "There is continuous extension whenever points are assumed to be so situated that there are no two between which there is not an intermediate point."
14. Cf. G.II.300: "Points are not parts of the continuum, but extremities, and there is no more a smallest part of a line than a smallest fraction of unity" (To Des Bosses). Both this and

the previous quotation are taken from Russell's excellent appendix of leading passages in his RuPL, pp. 247 and 248 respectively.

15. A further discussion of Leibniz's treatment of continuity, relating it to the concepts of combinatorial topology, is given in a paper by Graham Solomon and myself, "Leibniz's Conception of Continuity" (as yet unpublished).
16. The application of Extremum Principles in Physics, based on the simplicity of the underlying monads, is an example of what Leibniz called the Middle Science, the acquisition of knowledge of the phenomenal world by means of metaphysical principles of parsimony and optimization. For a discussion of this see Loemker's notes on pp. 27 and 61 of Loem.
17. Cf. also his letter to de Volder of 21/1/1704: "But all individual things are successive or subject to succession....Nor for me is there anything permanent in them other than that very law which involves a continued succession, the law in each one corresponding to that which exists in the whole universe" (G.II.263).
18. Evidently there is a slight ambiguity in Leibniz's usage of the term 'universe', in that it refers sometimes to the union of all possible worlds (U), and sometimes to the actual world (Wa). But it is always clear from context which meaning is intended.
19. See for instance Grunbaum's *Philosophical Problems of Space and time*, Dordrecht/Boston: D. Reidel, 1973, for an exposition of this theory.
20. Leibniz is quite explicit that this principle must be presupposed. Answering de Volder's objection that he derives the occurrence of changes from experience and does not show how it "follows from the nature of things itself", Leibniz replies: "Do you then think I either could or would want to demonstrate anything in nature if changes were not presupposed?...Nor have I tried to sell the idea [that changes emanate intrinsically [*ab intrinseco*]] on the basis of experience" (to de Volder, 21/2/1704: G.II.264).
21. For Leibniz a given structure is homogeneous if the whole is similar to any of its parts; this is avowedly true of time: "a temporal whole is similar to its part" (*Initia Rerum*, GM.VII.22), and this is the reason why space and time cannot

be real existents, as I shall explain in section 3. Cf. *Primae Veritates*, Cout.OF p 522: "For the diverse parts of an empty space would be perfectly similar and congruent to each other, and could not be distinguished from each other, and would therefore differ only in number, which is absurd. Time can also be proved not to be a thing, in the same way as space."

22. After writing this, I discovered that much the same point has already been made, and more cogently that here, by Fred D'Agostino in an article originally published in 1976 (*Philosophical Quarterly*, vol. 26, pp 125-138; reprinted in revised form in *Leibniz: Metaphysics and Philosophy of Science*, ed. R.S. Woolhouse, Oxford: Oxford UP, 1981, pp 89-103). D'Agostino argues that if the reducibility thesis is correct, "then all individual substances are compossible, and Leibniz's doctrine that there are many possible worlds becomes unsupportable", and consequently that Leibniz did indeed allow relational predicates to characterize individual substances.

23. My claim here that duration and extension considered in the abstract are the same as abstract time and space is a contentious one. In a weighty article that has recently been translated into English ("L'espace, le point et le vide chez Leibnz" *Revue philosophique de la France et de l'etranger* (1946) pp. 431-452, translated as "Space, Point and Void in Leibniz's Philosophy", pp 284-301 in LCIE), Martial Gueroult has argued for a profound distinction between space and abstract extension in Leibniz's conception: space, as an *innate idea*, "differs radically from the discursive concept of abstract extension (extensio) that is acquired by abstraction from a perceived property" (LCIE p 284). I confess that I cannot understand all Gueroult's arguments for this claim; although it is clear that Leibniz distinguishes space and extension inasmuch as not everything which is situated in space has extension, this only distinguishes extension *qua* attribute from space or abstract extension. The latter two terms, on the other hand, Leibniz appears to use interchangeably. Compare, for instance, his definitions in the *Initia Rerum* of duration and extension as "the magnitude of time" and "the magnitude of space", respectively, (GM.VII.18) with the claim in his

critique of Malebranche quoted in the text that time and space serve to measure duration and extension (considered as attributes of things) (G.VI.584).

24. This application of the term 'actual' to phenomena is something that seems to have escaped many commentators. It is of particular importance to recognize it in his treatment of the continuum problem if one is to make any sense of this, as I argue in my paper on Leibniz's conception of continuity (see note 4 above).
25. Clarke's actual argument was addressed to "imaginary space" (3rd Reply, #2, G.VII.368), and I have simply adapted it for the case of time.
26. In his influential work, *Syntagma Philosophicum*, Gassendi argued: "Thus we say that the world could have been created a thousand years before its creation, not because there were years marked off by the repeated revolutions of the sun, but because Time flowed of which the appropriate measures, the revolving motions of the sun, *could* have existed then. And we do not say that all these times were imaginary...." (Physicae Sectio I, Liber I, transl. Milic Capek and Walter Emge, in Capek p. 201).
27. "De Gravitatione et Aequipondio Fluidorum", pp. 75-156 in *Unpublished Scientific Papers of Isaac Newton*, ed. and transl. by A. Rupert Hall and Marie Boas Hall, Cambridge: Cambridge UP, 1962.
28. From the Scholium to the Definitions of Newton's *Mathematical Principles of Natural Philosophy* [1687], Cajori's 1934 transl., Capek p. 98.
29. Newton and Leibniz both believed that if only passive force, and not active force, were conserved, there would be a gradual running down of the total quantity of activity in the universe. Newton believed that God prevented this by his active intervention, a belief in keeping with the independence of time and activity. Leibniz believed that active force, his *vis viva* (mv^2), is conserved; if it weren't, the total activity in the unviverse could diminish to zero, in contradiction to the eternal activity of monadic appetition, of which phenomenal activity (or *vis viva*) is the result. With no appetition, of course, there would be no time according to Leibniz's theory.

In modern physics the connection between time and energy is explicit: according to Noether's Theorem, invariance under time translations is equivalent to conservation of energy.

30. As I have argued in the paper cited in footnote 15, it is also only abstract time that is ***truly continuous***, designating a possibility of division rather than a division into actual parts; whereas in concrete time, on the other hand, there are only determinate parts, there being more parts of the same kind in the greater time.
31. See the papers referred to in footnotes 4 and 15 above.
32. There is also the question of Leibniz's non-materialist conception of substance. From the beginning, simple substances were conceived by him as being ***principles*** of motion and continuation in existence, and thus as being more nearly ***mental*** or spiritual than material. For an excellent discussion of the interrelationship between the young Leibniz's metaphysics and theology and his physics of matter and motion, see Daniel Garber's "Motion and Metaphysics in the Young Leibniz", pp 160-184 in LCIE.
33. I have borrowed the image of the imaginary eye from Leibniz, in the *Phoranomus*, Cout.OF p 590.
34. It is worth noting that the connection among the states of a given monadic series makes for a fundamental disanalogy in Leibniz's mature system between spatial and temporal unities. For the unity of a perceiving being folows from a connection (*nexus*) among its states or perceptions: "The nexus of perceptions, according to which subsequent ones derived from preceding ones, gives rise to the unity of the percipient" (To des Bosses, 24/4/1709: G.II.372). But such a nexus is precisely what is lacking in an aggregate of monads (taken in themselves): "even monads by themselves do not compose a continuum, since they lack any connection (*nexus*) in themselves, any monad being just like a separate world" (To des Bosses, 29/5/1716: G.II.518). Thus the unity of a monad, which derives from the unique law which unites all its states into a series, is a real unity grounded in this law; it is a unity or identity of substance through time founded on the rational development of the universe; there is no analogous unity or identity of substance through space, and the spatial parts of

bodies do not compose into infinite true wholes, but only into phenomenal aggregates of states.

35. Again, see Garber's article cited in note 32 for a succinct account of Leibniz's early metaphysics of substance and motion.
36. See footnote 7 above.
37. See footnote 8 above.
38. See footnote 9 above.

William Seager

LEIBNIZ AND SCIENTIFIC REALISM

I want to enlist Leibniz's aid in answering a vexing question in the philosophy of science. It must be said at the outset that no one could confidently expect Leibniz to speak directly to any particular philosophical problem in this still young discipline. We will do better to search Leibniz for hints and suggestions -- and these we shall find aplenty.

A second preliminary point may be in order. I will discuss Leibniz's views on natural science, and especially his views on the *truth* of scientific theories. Thus I want to raise and dispose of the objection that Leibniz really thought that all scientific theories were false for the simple reason that they all dealt with the mere world of appearance or phenomena. These appearances all depend on the states of the monads, or *are* states of the monads (this is, of course, a complicated story), and only the monads actually exist. I suppose that this objection is literally correct, but stultifying. The distinction between appearance and reality is also drawn, by Leibniz, at the level of the monadic perceptions (see, for example, "On the Method of Distinguishing Real from Imaginary Phenomena".)[1] This is the level where science is at home -- and it is a level where the concepts of truth, falsity and evidence all apply.

The question at issue is simply whether we ought to believe scientific theories to be *true* or merely empirically adequate. Of course, such a bald and unqualified presentation of the question leaves it unanswerable. Presumably no one would hold that any extant scientific theory is either true or empirically adequate for the obvious reason that no extant theory enjoys unmitigated empirical success. But it's no false dilemma we have here -- only we need to make our question more precise.

Following van Fraassen let us say that a theory is empirically adequate

K. Okruhlik and J. R. Brown (eds.), The Natural Philosophy of Leibniz, 315–331.

if all *observable* phenomena fit inside some model of the theory, or if "what the theory says about the observable things and events in this world is true".[2] (This last is a quotation from van Fraassen's book, *The Scientific Image*, a recent and powerful attack on scientific realism.) Note that empirical adequacy extends across all time and space. If we drop the "observable" from the above, we arrive at what it is for a theory to be true simpliciter: not suprisingly, a theory is true simply if what it says about the world is true.

We still face the complicating difficulty that no extant theory is really deserving of our allegiance either with regard to truth or to empirical adequacy. But we may adopt van Fraassen's characterization of realism, and, as he calls his own position, constructive empiricism, to further clarify our question, which becomes: is it the *aim* of science to discover true theories or merely empirically adequate ones?

There would be no problem if theories restricted themselves to statements about the observable aspects of the world -- but of course they do not, and instead promiscuously entertain such dubious entities as atoms, humours, phlogiston, electrons, microbes, and on and on. Our problem thus becomes the question of whether science's employment of theories which mention unobservables amounts to the *postulation of the existence* of these things, or is merely a useful way to develop greater empirical adequacy.

The scientific realist is one who holds that when a scientific theory which mentions an unobservable is advanced, this advancing includes the positing of the existence of the unobservable.

I think there can be no doubt that Leibniz subscribed to scientific realism so construed. I believe this not because he ever actually says that, in general, the unobservables of scientific theory are to be taken as posits (though he may somewhere -- the Leibniz concordance might help here), but rather because of his free use of them in the construction of proto-explanations -- explanations which do not have the backing of any serious or well developed scientific theory. His "A New System of the Nature and the Communication of Substances, as well as the Union Between the Soul and the Body", for example, is one place, of many, where Leibniz puts forth his doctrine that at the death of an animal the "organic machine" is preserved as a home, as it were, for the soul. This, of course, is not observed by us -- but the explanation given is that the organic machine of the animal shrinks at death to "a size so small it escapes our senses just as it did before birth" (L p. 453). Perhaps in a more scientific vein we have Leibniz saying in "Reflections on the Common Concept of Justice":

> This is also true when one looks at the brain, which must undoubtedly be one of the greatest wonders of nature, since it contains the most immediate organs of sense. Yet one finds there only a confused mass in which nothing unusual appears but which nevertheless conceals some kind of filaments of a fineness much greater than that of a spider's web which are thought to be the vessels for that very subtle fluid called the animal spirits. Thus this mass of brain contains a very great multitude of passages -- and of passages too small for us to overcome the labyrinth with our eyes, whatever microscope we may use. For the subtlety of the spirits contained in these passages is equal to that of light rays themselves. Yet our eyes and our sense of touch show us nothing extraordinary in the appearnace of the brain. (L p. 565)

(Note here how Leibniz gives an example of unobservability which is based on theory -- a theory of light and vision.) Or again, in "On the Elements of Natural Science" Leibniz proposes that the explanation of the general operation of the human body, which is, after all, nothing but an "hydraulic-pneumatic machine" is in terms of certain "hidden ways ... in which composite things are dissolved into insensible parts" (L p. 283).

Leibniz is completely unselfconscious in his readiness to appeal to imperceptible aspects of nature, a readiness he shared with his contemporaries. Descartes, for example, explains rarefaction by appeal to insensible pores in bodies which can be opened to be filled by other matter, rather like a sponge (*Principles*, Pt. II, Principle 7) and Leibniz opined that Descartes had thus "admirably explained" rarefaction. In support of this mode of explanation Descartes says: "There is no compelling reason to believe that all bodies that exist must affect our senses" (*ibid.*)

Thus we are at least free to postulate their existence when necessary. Leibniz, in his criticism of the *Principles*, passes by this remark without comment -- it is too evident even to praise.

Here it my be worthwhile to forestall another objection stemming from Leibniz's overall metaphysics. Did he not hold that everything is perceived or represented in the perceptions of the monads (indeed, of any monad), and hence that there can be nothing which does not affect our senses? The reply, which I draw from Parkinson's "The Intellectualization of Appearances",[3] involves the distinction between distinct and confused perceptions. Everything is perceived only in the sense of confused perception, which is not sensation. When a perception becomes distinct it becomes a sensation (Leibniz says: "If a perception is more distinct it makes a sensation").[4] Since sensations are necessary for observation, the notion of an unobservable is perfectly acceptable to Leibniz.

The belief in unobservables which can somehow account for the phenomena we actually observe is not of course peculiar to the seventeenth century. It is found everywhere. A fragment from Anaxagoras is nice; it reads: "appearances are a glimpse of the unseen". The bare proposition that there are unobservables which have something to do with appearances is found everywhere because it is so obviously true. Observe the dust motes in a shaft of sunlight from across a room. As you walk closer more and more appear to you. It is, to me at least, overwhelmingly plausible that these motes grade off in size until some are truly unobservable. I hope to show that the bare plausibility of the existence of unobservables is one of the basic underpinnings of scientific realism.

And this plausibility might be thought enough of itself to answer our original question, for given the legitimacy of postulating unobservables, and the claim that any theory which is really empirically adequate is very *likely* to be true the dilemma of our original question appears to disappear.

Yes -- but suppose that for any theory T there was a second theory T' inconsistent with T but which utterly agreed with T about all observable things and events. That is, what if, for any theory there was an *empirically equivalent* rival? This is a possibility, and so long as the test of scientific theorising is empirical success there appears to be no ground for choosing one theory over the other.

I will take a moment to illustrate this with a concrete example used by van Fraassen, one striking in its simplicity. Consider Newtonian mechanics and gravitational theory. Within this theory uniform motion within absolute space is undetectable. This observation generates a multitude of empirically equivalent theories, each one positing a different uniform velocity in absolute space for, say, the sun. That each such theory *is* empirically equivalent is almost obvious (see van Fraassen's discussion in chapter 3 of *The Scientific Image*) and each is clearly inconsistent with any other. If science aims at truth then science has stalled.

This conclusion is escapable only if certain non-empirical determinants of theory choice can be brought into play here. But again, if the aim of science is truth these non-empirical criteria will have to be *evidential*, that is, have a bearing on the probable truth or falsity of theories. After all, there is no question but that we shall choose that theory which is easiest to use from any empirically equivalent grouping, yet ease of use is not obviously linked to truth. Let us call a non-empirical but evidential criterion of theory choice a "metaphysical principle".

Leibniz's reply to van Fraassen's example illustrates the use of metaphysical principles to evaluate theories. In the correspondence with Clark Leibniz argues that space is not absolute on the grounds that, if it were, God could "cause the whole universe to move forward in a right line or any other line, without making otherwise any alteration in it" (L p. 688). This would be, however, to make a "chimerical supposition". Of course, it follows that the assumption that the *sun* has any particular velocity in absolute space would be equally chimerical. Leibniz advances two related arguments in support of this; one from the principle of the identity of indiscernibles, another from the principle of sufficient reason. As to the latter, God could have no *reason* to favour the sun with any particular velocity (including velocity zero) over any other, and hence could not. And further, since Newton's theory leads inescapably to a world in which God faces this impossible choice it is Newton's theory, and hence every one of the empirically equivalent theories considered above, which must be wrong.

It is tempting to generalize Leibniz's argument to all cases of empirical equivalence. Suppose T and T' are empirically equivalent theories. Then the state wherein T holds is indiscernible from that wherein T' holds, hence there is no sufficient reason for T holding rather than T', or vice versa. By the above reasoning, both T and T' must be false. This attempt fails to convince, of course, but for a reason which supports our main point. The crucial claim that the states wherein T and T' hold are indiscernible is equivocal. They are so only empirically. The argument will fail where there is a non-empirical discernibility which is determinative for a rational and moral creature. For it is then that God could, and would, choose between making T or T' true of the world. (Creating a world is *precisely* to make a theory true. In van Fraassen's terminology, God is a model builder.) From the human standpoint, a non-empirical but determinative discernibility is a criterion of theory choice which is non-empirical but evidential -- that is, a metaphysical principle.

For Leibniz the metaphysical principles, such as that of continuity, operate as a sort of a filter through which theories or hypotheses must pass. But this filter is not an accurate gauge of empirical adequacy, for Leibniz admits that a false hypothesis may nonetheless be useful at the empirical level (L p. 283-84). However, the filter does play a clearly evidential role, for with regard to the Cartesian laws of impact Leinbiz shows how the principle of continuity can be used so that "the truth in these matters can be outlined in advance, as it were, or as a kind of prelude..." (L. p. 402).[5]

No one is surprised to see Leibniz employing metaphysical principles in theory evaluation. For everyone knows the rationalists -- sitting in their armchairs, strenuously applying pure reason to the world and faultlessly deriving absurdity after absurdity. But our neutral definition reveals that the use of metaphysical principles is no less, or not much less anyway, common today. I here offer two examples which have figured prominently in the discussion of this question.

Example 1. Wide-spread in discussions of scientific realism is the so-called principle of the common cause which asserts that science is charged with explaining any positive correlation (or perhaps, any correlation at all) between objects or events by way of a common cause of both. So we explain the correlation between the readings of two independent thermometers by the obvious common cause. This is a natural pattern of explanation in everyday life about oberservable phenomena, but if applied in general will demand the postulation of non-observables. For without this postulation there is no explanation. The postulation of unobservables by a given theory will in some cases, then, determine theory choice among empirical equivalents. It is clearly a non-empirical determinant, yet it is also evidential, for to postulate an unobservable thing is to postulate the truth of statements asserting the existence of the unobservable; the theory which does not posit them will thus miss the truth. The principle of the common cause so applied is thus a metaphysical principle. In fact, I cannot pass it by without mentioning the obvious links between this principle and the principle of sufficient reason.

Example 2. Consider now the argument vigorously advanced by R. Boyd[6] in support of scientific realism. Science has been and continues to be remarkably successful at the empirical level. This fact calls for an explanation -- the best being simply that science is true, or, as Boyd more coyly puts it, approximately true. If science at a given time is approximately true then new theories and extensions of old theories which are "in the tradition" or relevantly similar to older theories are to that extent more plausible, more likely to be true than exotic rivals. Once again, this principle of tradition-bound plausibility provides a non-empirical but evidential criterion of theory selection -- it is what we call a metaphysical principle.

There are many other examples to be found in discussions of scientific realism, but these two are particularly important, not least because of their wide range of applicability. Although the existence of such principles is not in doubt, their status is. The anti-realist can take issue with them in various ways, always attacking their evidentiality. Van Frassen argues in *The Scientific Image* that, since current science rejects a full application of the principle (it leads to the dreaded "hidden variables" in Quantum Mechanics), the principle of the common cause is only a methodological maxim for theory production; the principle is reduced "from a regulative principle for all scientific activity to one of its tactical maxims".[7] Boyd's principle is undercut at the first step -- the impressive empirical success which science currently enjoys is *not* necessarily best explained by the "approximate truth" (a rather murky notion anyway) of theories. Here, van Fraassen suggests to us instead that a Darwinian explanation of scientific success may be just as plausible.

In fact, the anti-realist can point to a more or less continual undercutting of one metaphysical principle after another throughout the history of science. Leibniz's principles suffered this fate rather early on, and to be thoroughly scientific we ought to reject all such extra-empirical determinants of theories and with it reject the outmoded notion that science aims at anything more than empirical success. Van Fraassen charges that: "realist yearnings were born among the mistaken ideals of traditional metaphysics".[8]

The battle lines can be clearly drawn. First, the simple postulation of unobservables which in some way or other account for observable phenomena is acceptable to reason. More, it appears to be an obvious truth that some such entities exist. (The doctrine of "scientific entity realism" as it is sometimes called may stop here, but if so it is not a very bold position.) If this were not so there would never have been a problem of scientific realism, for without such postulation empirical adequacy and truth never diverge. The anti-realist counters with the claim that for any theory there are empirically equivalent and incompatible rivals. This by itself does not show that the aim of science is not truth but rather empirical adequacy. But if one also supposes that the sole evidential criterion of theory choice is empirical success this conclusion is reasonable. For if we have a vast number of scientists spending their time constructing theories which account for and predict observable things and events (and it *does* seem that they spend very little of their time searching out empirical equivalents to pre-existing theories -- that is apparently reserved for philosophers), and the sole measure of success is just how well their theories do produce such accounts and predictions it is hard to resist the inference that their *aim* is the production of these accounts and predictions.

So we see why the realist attaches such importance to the metaphysical principles. I also believe that Leibniz was aware of this particular problem, as evidenced by his cautious endorsement of the so-called "conjectural method a priori" (see L. p. 283). This method is essentially the hypothetico-deductive method now much praised, but with regard to it Leibniz warns that "the same effect may have several causes". It is this very weakness that prompts Leibniz to relegate this method to "stop-gap" status -- useful hypotheses generated in this way are merely "to be presented, in the interim, in place of the true causes". The degree of probability these hypotheses can attain is apparently very great though -- Leibniz says they may be "morally certain" for example. He may be saying no more here than that such hypotheses cannot be *demonstrated,* which is, after all, true of all our scientific knowledge until we arrive in the "better life".

In any event, to carry out his attack, the anti-realist must convince us that (1) for any theory there are empirically equivalent rivals and (2) extra-empirical criteria of theory choice are non-evidential. Further, as I shall argue below, even if (1) and (2) can be supported they do not necessarily lead to anti-realism, for (1) can be construed in a weaker and a stronger way -- the anti-realist may require the strong sense for his anti-realist argument but the weak sense may be the only plausible form of (1).

The central issues that arise here obviously involve the status of methods of comparison among competing theories. I want to investigate these issues from a standpoint which provides an abstract but intuitively appealing picture of a field of competing theories. A particularly striking picture can be borrowed from computer science: it stems from a problem solving technique, especially used in artificial intelligence, called "hill-climbing".

The technique is quite nicely illustrated by the task of tuning into a television station. The measure of success is the quality of the picture; the means to achieve success are, let us say, adjustments of the tuning control and the brightness control. Abstractly speaking, we have a problem space of three dimensions -- one dimension for each parameter mentioned above. The plane formed by two adjustment axes will be dotted with hills representing the available channels. The hill-climbing technique involves moving through the problem space until one arrives at a "peak" -- a position where the associated control adjustments produce the best possible picture. This can be roughly illustrated so:

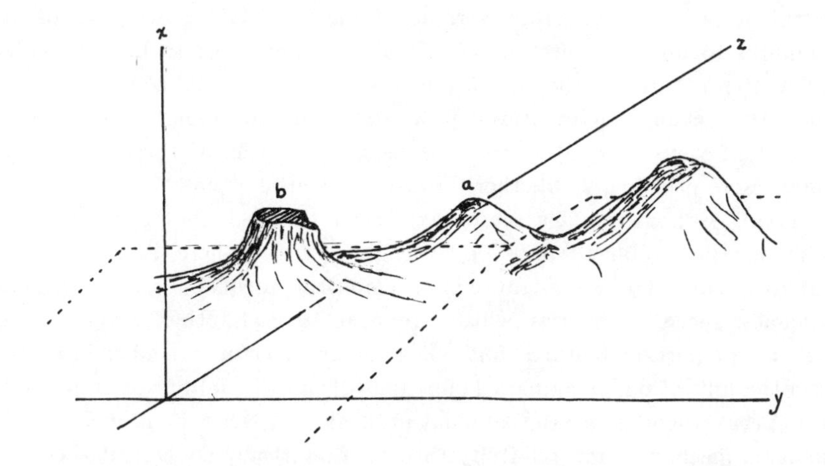

We can adopt this abstract picture and provide an encompassing view of competing theories. What is the measure of success of a theory? It is emprical adequacy. Both realist and anti-realist can agree on this, for while they disagree about the ultimate aim of science it is clear that the number one *measure* of success is the theory's ability to account for observable phenomena.

Thus we shall have empirical adequacy as the "up" axis which conventionally represents the measure of success. Needless to say, there is no presumption that this is the only criterion of scientific success. I do assume here, however, that theories can be graded as to how empirically adequate they are. This seems reasonable -- the notion of how near an empirical prediction is to the truth is clear enough for this purpose, and can often be quantified. (We hear scientists talk, for example, of a measurement being within two standard deviations of the predicted value.)

The other axes are much more difficult to label. A first point is that there is no need to restrict them to two -- the hill-climbing concept can be applied to a space of any dimension. The other axes can be simply *described* as features of theories which serve to differentiate them. Such features are difficult to isolate. Simplicity might be one such feature, whatever simplicity of theories *is*. In general, the metaphysical principles themselves will usually mark out a possible axis of comparison. The principle of continuity, for example, indicates an axis whereon theories are differentiated by whether they permit "nature to move by leaps", as it were, or how much of this they permit, or in what aspect of theory they permit it, and so on. However, as we are not trying to program a computer to produce our theories for us, we can afford to be unspecific about the nature of the various comparison axes. It is not a very contentious claim that theories can be compared one to another in a large number of ways -- what is contentious is the status of these modes of comparison, principally, whether they are evidential or not.

Having established this set of axes such theory takes its place as a point in the problem space -- since the problem here is to acheive empirical adequacy we might call it empirical adequacy space, or just adequacy space. Theories which are near to each other vis-à-vis the various comparison features but which differ in empirical adequacy will form the hills of our imaginary landscape -- from the unimpressive mound of impetus theories to the himalayan peaks of Newtonian mechanics, quantum mechanics and relativity theory. Empirically equivalent theories will lie on a plane cutting the axis representing empirical adequacy. The steeper the hill around a theory the fewer competitive rivals of the same ilk. An absolutely unique theory remains an isolated point in adequacy space with no geography around it.

Certain theses drawn from the philosophy of science can find a place in this picture. Recall that the notion of hill-climbing is the problem solving technique associated with an abstract problem space. In computer applications, the problem space is rigorously defined and the computer proceeds by taking what is called a standard step at each stage of the search for a solution. After each step, the "altitude" is measured, and the direction of the next standard step is varied in just the way you would expect. Now, we might use this progression by cautious steps as a metaphor for Kuhn's normal science. Hill-climbing fails in terrain like that at A (on the diagram), for every standard step leads down, yet A is only a local peak. Perhaps this is the situation when the procedures of normal science can no longer successfully attack the problems nature is posing for a given theory or range of theories. Scienfic revolutions are then the mounting of an expedition up an altogether new peak -- though the mechanisms by which this is accomplished are far from clear of course.

This picture can also be used to reinforce van Fraassen's rejection of Boyd's view that when a science reaches a certain level of empirical success it is then approximately true. Since deeply incompatible theories cannot all be approximately true (on any reasonable construal of "approximately") this amounts to a claim about the geography of adequacy space: namely that there is a level of empirical adequacy above which there is only one peak (the peak need not be too steep; for presumably similar, even if incompatible, theories can all be approximately true -- the various versions of general relativity might be an example of this). Boyd's argument is that only the approximate truth of a science can explain its empirical success, but this seems to beg the question since it is cogent only if the geography is just as he needs to argue for; and no independent reasons for mapping out adequacy space in this way have been given.

But the diagram also suggests a problem about the anti-realist thesis (1) -- that every theory has empirically equivalent rivals. Consider the region B. This plateau represents a number of quite similar and empirically equivalent theories, and can serve to represent the infinitely many Newtonian theories, discussed above, which differ only in the postulated absolute velocity of the sun. It is possible that all empirically equivalent theories of any degree of empirical success fall on such plateaus. (This is so because of the gappiness which is possible in adequacy space -- theories, or groups of theories can float in it.) If so, we have only a very weak empirical equivalence thesis and, I think, one which will not support the anti-realist position. For all such theories say *essentially* the same thing about the world. And they are just what the realist ought to expect given that there are elements of his theory which are supposed to be unobservable. In such a case it would be acceptable for the realist to admit that one could never know, say, the sun's absolute velocity but that this limitation does not impugn the general Newtonian world picture. Whichever theory is selected it leaves us with essentially the same world.

In this example the picture of the world which is common to them all is that theory in which the statement, "the velocity of the sun is v" is replaced by the statement "there is a velocity, greater than or equal to zero which the sun has".

Now, whether or not empirically equivalent theories always range themselves onto plateaus I can't say. Van Fraassen refers to Ellis's version of Newtonian cosmology in which no gravitational force is postulated but in which all motions are the same (the laws of motion incorporate gravitation), and Ellis's theory does not really seem very similar to Newton's.

Of course, once a theory exists it may be easy to produce empirically equivalent rivals to it. Is producing rivals in such a way some kind of trickery? As a sort of limiting case van Fraassen provides a method for producing any theory's empirically equivalent rival. Suppose we have a theory T. Then we can produce the rival thus: let T' be the theory which asserts that T is empirically adequate. Where is this theory to be placed in adequacy space?

Leaving the anti-realist thesis (1) for the moment let us consider the action of the metaphysical principles within adequacy space. The metaphysical principles restrict the range of acceptable theories, but without regard to empirical adequacy. Thus, for at least some of the principles, they outline a subplane of the theory comparison plane which is a sort of no-man's land. Theories in this region are not to be believed (we may, of course, use them for prediction). [Examples indicated on diagram]

Not every metaphysical principle operates in this way. For example, the principle of sufficient reason is supposed to rule out terrain like that at B but the principle does not clearly mark out a subspace. But a related principle perhaps does have an associated subspace. Some theories explain more than others, and such explanatory power is a virtue (though, again, it may not be an evidential virtue). We might suppose that some theories are to be rejected if they leave too much unexplained. Of course, it is on some such principle that the van Fraassen empirical equivalent of a theory might be rejected -- for the theory T' which says only that T is empirically adequate leaves much unexplained that T presumably can explain, though only in terms of unobservables.

The modern metaphysical principle which most clearly embodies this demand for explanatory power is that of the common cause. In *The Scientific Image*, van Fraassen shows that current science does not hold that this principle must have universal application. That is to say, current science does not hold that every correlation must be explained via a common cause. Note, it is clearly not just *empirical* correlations that are to be explained by the principle -- for in the electron spin cases, for example, we *can* explain the empirical correlation of the particle detector states, and Quantum Theory does so in terms of the states of unobservables which are subject to an isomorphic correlation. According to van Fraassen this limitation on the principle destroys any presumption that it is evidential and reduces it to a mere "tactical maxim".

Now, I think it is *possible* to imagine a theory which could apply this principle without limit and explain everything. This could be done in a theory which gave micro-explanations of any correlation, where these explanations were produced according to some recursively defined procedure. Theorists could then sequentially explain features of the world by appeal to ever smaller entities or ever more hidden explanatory factors. Everything could be explained.

The demand for the complete applicability of the principle of explanatory power can be represented on the diagram. If one axis of comparison is explanatory power then the entire sub-plane is to be marked out of bounds up to the unit of *full* explanatory power. This is, thus, an extremely stringent demand on theories.

The alternative is to leave some things unexplained -- to leave some of the so-called "cosmic coincidences" in one's theory. Van Fraassen claims that this is to admit the tenability of the anti-realist position. But surely there are *grades* of cosmic coincidences. A theory which posits only a few sorts of coincidences -- a few sorts of brute facts is certainly *better* than a theory which leaves us with a vast number of brute facts. (Especially if the residual brute facts are about what the theory itself proclaims to be very simple sorts of things, like, needless to say, electrons.) Is it more likely to be true? It might well seem so, because it is initially likely that observable phenomena are in some way the result of hidden activity. And also because it is initially plausible that correlations can be explained. If there are some truly brute facts this is perhaps unfortunate from the point of view of a curious species, but it does not seem impossible. Still, one should be *forced* to accept any fact as brute. The van Fraassen empirical equivalent of a theory T leaves us with so many brute facts which we are not forced to accept, since we have the theory T itself to explain most of them, that this empiricial equivalent can be rejected relative to T.

Van Fraassen has a strikingly simple reply to this sort of argument.

One ought to believe the more probable of two theories, if one is to believe any at all. The van Fraassen empirical equivalent T' of a theory T is related to T in the following way: T = T' + E, where E is the set of statements which assert the existence of the various unobservables mentioned by T. Obviously, then, T' cannot have a lower probability than T. Indeed, so long as E is not certain, T' will inevitably be more probable than T.

The way I am looking at this problem, there appears to be something wrong with this argument. The van Fraassen empirical equivalent T' is supposed to entail that there are many sorts of brute facts, but if T = T' + E then T must entail that there is precisely the same set of brute facts. This does not seem right. Brute facts are facts which are not explained by a theory. For example, the fact that pressure increases with temperature (given fixed volume) is a brute fact on the van Fraassen empirical equivalent of the kinetic theory of gases. It is not a brute fact on the kinetic theory itself. So here T does not seem to be just T' + E.

The tension here might be explained in terms of some kind of non-standard implication relation (the counter-factual implication has the feature that T's implying B does not entail that T + X implies B for example), or, better, by noting the intensional character of brute facts which stems from the intensional context "e explains...". But another source of tension is the question of the *rivalry* between T and T'. If they are construed as being incompatible then one cannot be a logical strengthening of the other (at least if both are consistent). And there is a natural tendency toward such a view. The anti-realist *could* be claiming that the world only has observable features and that his theory is the whole story about these. (In fact, van Fraassen admits that the anti-realist must accept many facts as brute which are explicable on a realist theory.[9]) Then T is considered to be false by the anti-realist. It is this anti-realist position we can reject by the above argument. A weaker anti-realist position is a form of agnosticism. It says merely that we will never *know* that any theory which mentions hidden structure is true. But equally we shall never know that any theory is empirically adequate. The agnostic anti-realist nevertheless will countenance the acceptance of a theory as empirically adequate in the right evidential circumstances. The difference in attitude must lie in a difference in the support the evidence gives for the two sorts of theory. The van Fraassen empirical equivalent of T is supported by any evidence that supports T itself, and vice versa.

Where is the difference? The difference is that the van Fraassen empirical equivalent of T is a priori more probable than, or at least as probable as, T. That this does not settle the question is obvious from the consideration that nonetheless the probability of T may be very high. Since within an agnostic anti-realism the van Fraassen empirical equivalent of T is not taken to be a rival of T, an increase in its probability has no downward effect on that of T.

To finally draw Leibniz back into the discussion, consider a principle he puts forth: "it must be admitted that a hypothesis becomes the more probable as it is simpler to understand and wider in force and power, that is, the greater the number of phenomena that can be explained by it.... It may even turn out that a certain hypothesis can be accepted as physically certain if, namely, it completely satisfies all the phenomena which occur, as does the key to a cryptograph" (L p. 188). The image of the crypotgraphy is really perfect here, for as Leibniz admits, it is at least often possible to find another key which makes sense of the encrypted message (see L p. 283).

The probability of a theory is then dependent on the number of incompatible and significantly different empirical equivalents to it. No one knows whether every theory has such. In the absence of competitors the probability that a theory is true can reach levels where any reasonable person ought to believe it to be true.

But as observed at the very beginning no extant theory appears to have reached this degree of probability, despite the glaring lack of empirically equivalent rivals which an anti-realist might invoke to draw down a given theory's likelihood. So let me close by considering the status of science in the real situation of no unqualified empirical success and a presumption of theory replacement uniformly borne out by the history of science. Throughout this history the gauge of success has been empirical adequacy. But equally pervasive has been the presence of theories which postulate hidden structures in nature. Each of these theories is false, but they all imply a truth: there is hidden structure. In this they clearly do better than the van Fraassen equivalents which merely assert that each theory is empirically adequate. Nature *is* subtle. The postulation is exactly what we would exect of someone trying to fathom this subtlety; it is not what we would expect of those seeking only empirical adequacy. It is because nature has this hidden subtlety that only theories which posit hidden structure have any hope of empirical adequacy. It is reasonable to suppose, then, that scientists are attempting to grapple with these hidden features and force them into the open. This argument is from Leibniz, and is expressed in a beautiful passage:

> Unless principles are advanced from geometry and mechanics which can be applied with equal ease to sensible and insensible things alike, nature in its subtlety will escape us. And reason must supply this most important lack in experiment. For a corpuscle hundreds of thousands of times smaller than any bit of dust which flies through the air, together with other corpuscles of the same subtlety, can be dealt with by reason as easily as can a ball by the hand of a player (L p. 283).

Further, we note that the theories which reason presents to us dealing with various aspects of nature are all under a constraint of compatibility. Science rests uneasy whenever powerful theories seem not to go together. There is no need for this unease if empirical adequacy is the object of the exercise. But if, for example, genetics cannot be integrated into physics in some way or other something has gone wrong, even if both are empirically successful within their own domains. The reply might be made that only by seeking ever more encompassing theories will true empirical adequacy be achieved. It is true that science seems to seek "total theories", theories in which every question about nature has an answer. But this cannot be just the quest for empirical adequacy for this quest cannot result in a total theory. Such a theory would assert "there is no hidden structure". It is then a true rival to those theories which do posit hidden structure, and, as we saw above, is extremely implausible and can be rejected. I conclude that the urge to unification in science is a sign that it is the truth which is being sought and not just empirical adequacy.

Department of Philosophy
University of Toronto

Notes

1. In L. Loemker (ed.) *Philosophical Papers and Letters,* Reidel, 1956, p. 363. Henceforth I will refer to the Loemker volume as "L".
2. Van Fraassen, B. *The Scientific Image.* Clarendon Press, Oxford, 1980, p. 12.
3. In M. Hooker (ed.) *Leibniz: Critical and Interpretive Essays.* University of Minnesota Press, 1982, pp. 3-20.
4. G.H.R. Parkinson, ed. *Philosophical Writings,* p. 85.
5. For more on the general operation of the metaphysical principles see my "The Principles of Continuity and the Evaluation of Theories", *Dialogue,* vol. 20, 3, 1981.
6. See R. Boyd, "Realism, Underdetermination and a Causal Theory of Evidence" in *Nous,* March 1973 and "Scientific Realism and Naturalistic Epistemology" in *Proceedings of the 1980 Biennial Meeting of the Philosophy of Science Association, Volume Two* (ed. P. Asquith and R. Giere), Philosophy of Science Association, 1981.
7. *The Scientific Image,* p. 31.
8. *Ibid,* p. 23.
9. *Ibid,* p. 24.

INDEX

(*Compiled by Gordon McOuat*)